STUDENT STUDY GUIDE

Geralyn M. Koeberlein
Mahomet-Seymour High School

Daniel C. Alexander
Parkland College

Christine S. Verity

ELEMENTARY GEOMETRY
for College Students

Fourth Edition

Daniel C. Alexander
Parkland College

Geralyn M. Koeberlein
Mahomet-Seymour High School

HOUGHTON MIFFLIN COMPANY   BOSTON NEW YORK

Publisher: Richard Stratton
Senior Sponsoring Editor: Lynn Cox
Associate Editor: Melissa Parkin
Assistant Editor: Noel Kamm
Editorial Assistant: Laura Ricci
Project Editor: Carol Merrigan
Editorial Assistant: Eric Moore
Manufacturing Coordinator: Renee Ostrowski
Senior Marketing Manager: Katherine Greig
Marketing Assistant: Naveen Hariprasad

Printed in the U.S.A.

ISBN 13: 978-0-618-64526-8
ISBN 10: 0-618-64526-8

1 2 3 4 5 6 7 8 9-VGI-10 09 08 07 06

Contents

How to Study Geometry

Textbook:
1. Read the textbook word-for-word.
2. Pay special attention to undefined terms. Later textbook topics depend upon these terms.
3. When you encounter a new vocabulary term:
   a. Make a drawing that illustrates the concept.
   b. Think of an example that uses the concept.
   c. State the definition in terms that work for you.
4. When you encounter a new postulate or theorem:
   a. Read/reread until you understand the statement.
   b. Make a drawing to illustrate the statement.
   c. State the postulate/theorem in your own words, without changing its meaning.

NOTE: The student may wish to create an index card to correspond to each important term, postulate, or theorem. On each card, state the definition, postulate, or theorem; illustrate with a drawing and a pertinent example.

5. When you encounter example:
   a. Read it step-by-step, justifying each conclusion as the example unfolds.
   b. Refer to drawings that provide visual support for the steps of the example.
   c. With the textbook closed, try repeating the steps of the example.
6. When you encounter a completed proof (like in an example):
   a. Note the order of statements and the reasons that justify these claims.
   b. Reason from the given information by studying the related drawing.
   c. Consider statements in reverse order, noting that the final statement should be the "Prove" statement. Ask "What previous statement would allow the conclusion found in the following statement?"
   d. Try repeating the statements and reasons of the proof with the textbook closed.
7. Classify by groups; for instance, write lists of methods that verify that "lines are parallel," "triangles are congruent," and so on.

Assignments:
1. Complete as many of the assigned problems as possible. Do as much as you can do on your own! Use the textbook to find the principle (definition, postulate, theorem) that provides background for the problem.
2. You may eventually need to work with a study buddy or seek tutorial help by attending the mathematics laboratory (if one is available).
3. Use the Student Study Guide when you are stuck. Use it to help you complete the problem/proof. Do NOT simply read the solution, because that does not help the student generate a solution!
4. When an assigned problem seems difficult or impossible, go on to the next problem. In some cases, the solution for one problem will provide insight into the solution for another.

Preparing for a Test:
1. Prepare a list of the important concepts found in the textbook material upon which the test is based.
2. Study chapter summaries, review exercises, and try the practice test found at the end of chapters that are within the test material.
3. Review assignment problems that caused difficulty at the time of the assignment.
4. If confused, consult the instructor, a tutor, a study buddy, or a lab monitor.
5. Study with a study group if you tend to learn while working in a cooperative setting.

Attendance:
1. Attend all classes! Arrive on time! If you must be absent, ask the instructor if your school has videos that accompany the textbook
2. Staying informed/keeping current provide you with the knowledge to make geometry interesting.

Section-by-Section Objectives

The following are the student goals for *Elementary Geometry for College Students*, Fourth Edition by Alexander and Koberlein.

Chapter One – Line and Angle Relationships

Section 1.1: Statements and Reasoning
1. Determine whether a collection of words forms a statement
2. Form the negation of a given statement
3. Form the conjunction, disjunction, or implication determined by two simple statements
4. Recognize the hypothesis/conclusion of a conditional statement
5. State and recognize the three types of reasoning used in geometry
6. Recognize/apply the Law of Detachment
7. Form the intersection, union of sets
8. Illustrate a deductive argument visually by using a Venn Diagram

Section 1.2: Informal Geometry and Measurement
1. Describe the terms point, line, and plane
2. Become familiar with geometric terms such as collinear, line segment, and angle
3. Measure a line segment with a ruler and/or measure an angle with a protractor
4. Write equations based upon statements involving midpoint, bisect, and congruent
5. Recognize the terms right angle, straight angle, and perpendicular
6. Use the compass to construct a line segment of specified length
7. Use the compass to determine the midpoint of a given line segment

Section 1.3: Early Definitions and Postulates
1. State the four parts of a mathematical system: undefined terms, definitions, postulates, and theorems
2. Recognize the need for/characteristics of a precise definition
3. Write symbols for line, ray, line segment, and length of line segment
4. State the initial postulates for lines and planes (in your own words)
5. Apply the Segment-Addition Postulate
6. Recognize the concepts parallel lines/parallel planes

Section 1.4: Angles and their Relationships
1. Write the symbols that name an angle or the measure of that angle
2. Understand/use terms related to angles (like sides, vertex, etc.)
3. State/apply postulates involving an angles(s)
4. Given an angle's measure, classify the angle as acute, right, obtuse, straight, or reflex
5. Apply the Angle-Addition Postulate
6. Classify pairs of angles as adjacent, congruent, complementary, etc
7. Use the compass to construct an angle congruent to a given angle
8. Use the compass to construct the bisector of a given angle

Section 1.5: Introduction to Geometric Proof
1. Demonstrate the two-column form of a proof
2. Understand the role of the Given, Prove, and Drawing for a proof problem
3. Provide reasons that justify statements found in proofs
4. Provide statements that are justified by the reasons found in proofs

Section 1.6: Relationships – Perpendicular Lines
1. Know/apply the definition of perpendicular lines
2. Recognize the general concept "relation"
3. Understand/apply the reflexive, symmetric, and transitive properties of congruence
4. At a point on a line, construct the line perpendicular to the line
5. Construct the perpendicular-bisector of a given line segment

Segment 1.7: The Formal Proof of a Theorum
1. Determine the hypothesis and conclusion of a given theorem
2. State the five written parts of the formal proof of a theorem
3. Make a "Drawing" that is suitable for the proof of a theorem and which is based upon the hypothesis of the theorem
4. Write the "Given" for the proof of a theorem, based upon the hypothesis of the theorem and a suitable "Drawing"
5. Write the "Prove" for the proof of a theorem, based upon the conclusion of the theorem and a suitable "Drawing"
6. State/apply theorems involving perpendicular lines, complementary angles, and so on
7. Construct/complete the formal proof of a theorem

Chapter Two – Parallel Lines

Section 2.1: The Parallel Postulate and Special Angles
1. From a point not on a line, construct the line perpendicular to the given line
2. Recognize when two lines, a line and a plane, or two planes are perpendicular
3. Recognize when two lines, a line and a plane, or two planes are parallel
4. Define parallel lines and parallel planes
5. Understand/apply terms such as transversal, corresponding angles, etc.
6. State/apply initial postulates involving parallel lines
7. State/apply selected theorems involving given parallel lines

Section 2.2: Indirect Proof
1. Know the true/false relationships between a conditional statement and its converse, inverse, and contrapositive
2. State/apply the Law of Negative Inference
3. State/apply the method of indirect proof
4. Know that negations and uniqueness theorems are often proved indirectly

Section 2.3: Proving Lines Parallel
1. State/apply/prove selected theorems establishing that lines are parallel
2. From a point not on a line, construct the line parallel to the given line

Section 2.4: The Angles of a Triangle
1. Know definitions of triangle and related terms (vertices, sides, etc.)
2. Classify triangles by side relationships (scalene, isosceles, equilateral)
3. Classify triangles by angle relationships (acute, right, obtuse, and equiangular)
4. Apply the theorem, "The sum of angles of a triangle is 180°"
5. State/apply/prove the corollaries of the theorem stated in (4)

Section 2.5: Convex Polygons
1. Know the definitions of polygon and related terms

2. Classify polygons as convex/concave and by their numbers of sides
3. Determine the number of diagonals for a polygon of *n* sides
4. State/apply theorems involving sums of angle measures of a polygon
5. Classify polygons as equiangular/equilateral/regular
6. Recognize a figure that is a polygram/regular polygram

Section 2.6: Symmetry and Transformations
1. Recognize whether a figure has one or more lines of symmetry
2. Classify line symmetry as vertical, horizontal, or neither
3. Determine whether a figure has point symmetry
4. Recognize/classify/draw transformations known as slides, reflections, or rotations
5. Know that every transformation of a figure produces an image that is congruent to the given figure

Chapter Three – Triangles

Section 3.1: Congruent Triangles
1. State the definition of congruent triangles
2. Determine the correspondences between parts of congruent triangles
3. Name the included side (angle) for 2 angles (2 sides) of a triangle
4. Know/determine/apply the methods (SSS, SAS, ASA, and AAS) for proving triangles congruent

Section 3.2: Corresponding Parts of Congruent Triangles
1. Know that/use CPCTC to symbolize "Corresponding parts of congruent triangles are congruent"
2. Recognize the types of conclusions that can be established by using CPCTC
3. Use markings on congruent triangles to indicate their corresponding parts
4. State/apply the HL theorem
5. State/apply the Pythagorean Theorem
6. Determine the method that establishes that triangles are congruent

Section 3.3: Isosceles Triangles
1. In a triangle, distinguish between angle-bisector of angle, altitude, perpendicular-bisector of side, and median
2. Know that a triangle has three angle-bisectors, altitudes, medians, and perpendicular-bisectors of its sides
3. Decide whether the description of an auxiliary line is determined, overdetermined, or underdetermined
4. State/use "If two sides of a triangle are congruent, the angles opposite these sides are congruent" and its converse
5. State/apply the definition of perimeter of a triangle

Section 3.4: Basic Constructions Justified
1. Construct/justify the construction of an angle congruent to a given angle
2. Construct/justify the angle-bisection method
3. Construct/validate the construction of line segments of specified length
4. Construct/validate the construction of angles of specified measure
5. Construct/validate the construction of selected regular polygons

Section 3.5: Inequalities in a Triangle
1. Know/apply the definition of "is less than"
2. Use the relationships found in lemma (helping theorems) of this section
3. State/apply theorems involving inequalities in a triangle
4. State/apply corollaries involving the length of a line segment from a point not on a line (or plane) perpendicular to the line or plane
5. State/apply the Triangle Inequality (or alternate form)

Chapter 4 – Quadilaterals

Section 4.1: Properties of a Parallelogram
1. State definitions for quadrilateral and parallelogram
2. State/apply/prove selected theorems involving given parallelograms
3. Use angle measures of a parallelogram to determine its longer/shorter diagonal
4. Determine speed/direction of a moving airplane subject to wind conditions

Section 4.2: The Parallelogram and Kite
1. Know that the parallelogram/kite each have two pairs of congruent sides
2. Know that quadrilaterals with congruent opposite sides are parallelograms
3. Know that quadrilaterals with a pair of congruent and parallel sides are parallelograms
4. Know that quadrilaterals with diagonals that bisect each other are parallelograms
5. Know that the kite has one pair of opposite angles that are congruent
6. Know that the kite has a diagonal that is the perpendicular-bisector of the other diagonal
7. State/apply the theorem in which the midpoints of two sides of a triangle are joined

Section 4.3: The Rectangle, Square, and Rhombus
1. State definitions for the rectangle, square, and rhombus
2. State/apply/prove theorems involving the rectangle/square/rhombus
3. State/apply/prove corollaries involving the rectangle/square/rhombus
4. Apply the Pythagorean Theorem to quadrilaterals

Section 4.4: The Trapezoid
1. Know the terminology related to the trapezoid and isosceles trapezoid
2. State/apply/prove theorems and corollaries involving trapezoids
3. State/apply the theorem "If three (or more) parallel lines intercept congruent segments on one transversal, then they intercept congruent segments on any transversal."

Chapter Five – Similar Triangles

Section 5.1: Rations, Rates, and Proportions
1. State/apply the terms ratio, rate, and proportion
2. Know the terminology (means, geometric mean, etc.) related to proportions
3. State/apply the Means-Extremes Property (of a proportion)
4. Understand/apply further properties of proportions

Section 5.2: Similar Polygons and Triangles
1. Form an intuitive understanding of the concept "similarity of figures"
2. Determine the correspondences between the parts of similar polygons

3. Utilize correspondence relationships to find angle measures and lengths of sides of similar polygons

Section 5.3: Proving Triangles Similar
1. Name 3 methods (AA, SAS~, and SSS~) used in proving triangles similar
2. Choose a method and then prove that two triangles are similar
3. Use CASTC to prove that angles are congruent (corresponding angles of similar triangles are congruent)
4. Use CSSTP to prove a proportion involving lengths of sides of similar triangles (corresponding sides of similar triangles are proportional)

Section 5.4: The Pythagorean Theorem
1. State/prove/apply Theorem 5.5.1 in establishing later theorems
2. State/apply/prove theorems involving geometric means in the right triangle
3. State/apply the Pythagorean Theorem and its converse
4. Determine whether (a,b,c) is a Pythagorean Triple
5. Determine whether a triangle is acute, right, or obtuse based upon the lengths of the sides of the triangle

Section 5.5: Special Right Triangles
1. State/apply/prove the 45-45-90 Theorem
2. State/apply/prove the 30-60-90 Theorem
3. Recognize/apply the equivalent theorems such as Theorems 5.5.3 and 5.5.4

Section 5.6: Segments Divided Proportionally
1. Form an intuitive understanding of the concept "segments divided proportionally"
2. State/apply the definition of segments divided proportionally
3. State/apply/prove the theorem that establishes that parallel lines determine proportional segments on transversals
4. State/apply the theorem, "The angle-bisector in a triangle separates the opposite side of the triangle into segments whose lengths have the same ratio as the lengths of the sides of the bisected angle"
5. Use Ceva's Theorem to form an equation involving the lengths of the parts of the sides of a triangle

Chapter Six – Circles

Section 6.1: Circles and Related Segments and Angles
1. Become familiar with the terminology (radius, center, chord, arc, etc.) of the circle
2. State/apply postulates related to the circle
3. State/apply/prove selected theorems related to the circle
4. State/apply methods of measuring central/inscribed angles in the circle

Section 6.2: More Angle Measures in the Circle
1. State definitions for terms such as tangent and secant (of a circle)
2. Recognize when polygons are inscribed in/circumscribed about a circle
3. Recognize when circles are inscribed in/circumscribed about polygons
4. State/apply methods of measuring central/inscribed angles in the circle

Section 6.3: Line and Segment Relationships in the Circle
1. State/apply/prove theorems relating radii and chords of a circle
2. Recognize/use terminology involving tangent circles
3. Recognize/use terminology involving common tangents to circles
4. State/apply/prove theorems involving lengths of chords, tangents, and secants

Section 6.4: Some Constructions and Inequalities for the Circle
1. State/apply Theorem 6.4.1 (radius drawn to point of tangency is perpendicular to tangent)
2. Perform constructions of tangent to circle at a point on the circle or from a point in the exterior of the circle
3. State/apply prove theorems relating unequal chords, arcs, and central angles of a circle

Section 6.5: Locus of Points
1. Understand/state the definition of the term locus
2. Draw/construct/describe the locus of points for a selected condition(s)
3. Recognize the locus of points equidistant from sides of angle (and from endpoints of line segment)
4. Recognize/describe the difference between a locus in a plane/space
5. Verify the locus theorem by establishing two results

Section 6.6: Concurrence of Lines
1. Understand/state the definition of concurrent lines
2. State/apply/prove the concurrence of the three angle-bisectors of a triangle
3. State/apply/prove the concurrence of the perpendicular-bisectors of sides of a triangle
4. State/apply the concurrence of the three altitudes of a triangle
5. State/apply the concurrence of the three medians of a triangle
6. Know the meaning of the terms: incenter, circumcenter, orthocenter, and centroid of a triangle

Chapter Seven – Areas of Polygons and Circles

Section 7.1: Area and Initial Postulates
1. Develop an intuitive understanding of the area concept
2. Distinguish between units of length and units of area measurement
3. State/apply the initial postulates involving areas of regions
4. Prove/apply theorems involving area of a square, parallelogram, or triangle

Section 7.2: Perimeter and Area of Polygons
1. State/apply perimeter formulas for selected polygons
2. State/apply Heron's Formula for the area of a triangle
3. State/apply/prove formulas for areas of trapezoid, rhombus, and kite
4. Use the ratio between the lengths of corresponding sides of similar polygons to determine the ratio between their areas

Section 7.3: Regular Polygons and Area
1. Determine whether a given polygon can be inscribed in a circle
2. Determine whether a given polygon can be circumscribed about a circle
3. Perform constructions involving inscribed/circumscribed polygons in and around circles
4. Calculate measure of central angle, radius, and apothem of a regular polygon
5. Determine the area of a regular polygon by applying the formula $A = aP$

Section 7.4: The Circumference and Area of a Circle
1. Recall that π is the ratio of the circumference to the length of the diameter of a circle
2. Know/apply the formulas $C = d$ and $C = 2r$ for the circumference of a circle
3. Memorize the common approximations for π
4. Understand/apply the formula for the length of an arc
5. State/apply the formula $A = r$ for the area of a circle

Section 7.5: More Area Relationships in the Circle
1. Understand/apply the formula for the area of a sector
2. Determine the area of a segment of a circle
3. Prove that the area of a triangle with a perimeter P and radius r of inscribed circle is given by $A = rP$
4. Determine the area of a triangle using the formula $A = rP$

Chapter Eight – Surfaces and Solids

Section 8.1: Prisms, Area, and Volume
1. Understand intuitively the notion of prism
2. Understand/use terminology (edges, vertices, etc.) related to prisms
3. Determine the lateral area/total area of a prism
4. Memorize/apply the formula for the volume of a prism

Section 8.2: Pyramids, Area, and Volume
1. Understand intuitively the notion of a pyramid
2. Understand/use terminology (edges, vertices, etc.) related to pyramids
3. Apply $\ell = a^2 + h^2$, which relates slant height, apothem, and altitude of a regular pyramid
4. Apply $e^2 = h^2 + r^2$, which relates lateral edge, radius, and altitude of a regular pyramid
5. Determine the lateral area/total area of a pyramid
6. Memorize/apply the formula for the volume of a pyramid

Section 8.3: Cylinders and Cones
1. Understand intuitively the notions of cylinder and cone
2. Understand/use terminology related to cylinders and cones
3. Apply $\ell^2 = r^2 + h^2$, which relates slant height, radius, and altitude of a right circular cone
4. Memorize/apply formulas for lateral/total area of a right circular cylinder
5. Memorize/apply formulas for lateral/total area of a right circular cone
6. Memorize/apply formulas for the volume of a right circular cylinder/cone

Section 8.4: Polyhedrons and Spheres
1. Understand intuitively the notion of polyhedron
2. Know/use the terminology related to a polyhedron/regular polyhedron
3. Verify Euler's Equation, $V + F = E + 2$, relating the numbers of vertices, faces, and edges of a polyhedron
4. State/describe the five regular polyhedrons
5. Know/apply the term sphere and terminology related to a sphere
6. Memorize/apply formulas for surface area/volume of a sphere
7. Understand/apply the concept of solid of revolution

Chapter Nine – Analytic Geometry

Section 9.1: The Rectangular Coordinate System
1. Know/use terms related to the rectangular coordinate system
2. Plot/read points in the coordinate system as ordered pairs
3. Find the distance between two points on a vertical/horizontal segment
4. Know/apply/prove the Distance Formula
5. Know/apply the Midpoint Formula

Section 9.2: Graphs of Linear Equations and Slope
1. State/apply the definition of graph of equation
2. Determine/use intercepts in graphing straight lines
3. Know/apply the Slope Formula
4. Determine by sight if a line has positive/negative/zero/undefined slope
5. Use slope relationships to determine if lines are parallel/perpendicular

Section 9.3: Preparing to do Analytic Proofs
1. Determine the analytic formula necessary to prove a given statement
2. Prepare the drawing used to complete the analytic proof of a theorem
3. Name/describe general coordinates of vertices for a particular type of geometric figure
4. Use algebraic relationships to develop geometric relationships for given geometric figures

Section 9.4: Analytic Proofs
1. Develop a logical, orderly plan needed to complete an analytic proof
2. Construct the analytic proof of a given geometric theorem

Section 9.5: Equation of Lines
1. Use the Slop-Intercept and Point-Slope Forms to find equations of lines
2. Use the equation of a line to draw its graph
3. Use graphs/algebra to solve systems of equations (find points of intersection)
4. Develop analytic proofs for theorems by using equations of lines

Chapter Ten – Introduction to Trigonometry

Section 10.1: The Sine Ration and Applications
1. Define/apply the sine ration of an acute angle of a right triangle
2. Use a table/calculator to determine the sine ration of an acute angle
3. Use a table/calculator to determine the acute angle whose sine ratio is known
4. Understand/apply the notion angle of elevation/depression

Section 10.2: The Cosine Ration and Applications
1. Define/apply the cosine ration of an acute angle of a right triangle
2. Use a calculator to determine the cosine ration of an acute angle
3. Use a calculator to determine the measure of an acute angle whose cosine ratio is known
4. Know/apply/prove the identity $\sin^2 \theta + \cos^2 \theta = 1$

Section 10.3: The Tangent Ratio and Other Ratios
1. Define/apply the tangent ratio of an acute angle of a right triangle

2. Recognize which trigonometric ratio (sine, cosine, tangent) can be used to determine an unknown measure in a right triangle
3. Use a calculator to determine the tangent ratio of an acute angle of a right triangle
4. Use a calculator to determine the measure of an acute angle whose tangent ratio is known
5. State/apply the definitions of the cotangent, secant, and cosecant ratios for an acute angle of a right triangle
6. Define/determine $\cot\theta$, $\sec\theta$, and $\csc\theta$ as reciprocals of $\tan\theta$, $\cos\theta$, and $\sin\theta$, respectively

Section 10.4: More Trigonometric Relationships

1. State/apply the formula $A = \dfrac{1}{2}ab\sin\gamma$ (or equivalent)
2. State/apply the Law of Sines
3. State/apply the Law of Cosines
4. use given measure to decide whether the Law of Sines/Cosines should be used to find an unknown measure in a triangle

To the Student:

This interactive companion was prepared for students so that they could work with the instructor or fellow students to become familiar with the most important concepts of each section of the textbook <u>Elementary Geometry for College Students, 4e</u>. By working together with the instructor and/or classmates, the student will most likely be well aware of the terminology, symbols, principles, and techniques necessary to perform well in the Geometry class.

Following a class in which you have used the interactive companion, you (the student) will be better prepared to delve more deeply into the subject matter of Geometry. After each class, it is highly recommended that you read each section completely. Reading will enable you to expand upon the knowledge gained in the classroom experience; in general, there is simply not enough time in a classroom period to fully explore each topic. Having used the interactive companion in class, you will also be better prepared to do assigned problems of each section of the the textbook as homework. Although many of the textbook problems are intended to reinforce material presented in the classroom, the majority of the problems delve deeper and so are more difficult, more theoretical, or more applied than those of the interactive companion.

The interactive companion is *not intended* as an alternative to the homework generally assigned from the textbook. When used in conjunction with the textbook, it is likely that the student's level of proficiency in Geometry will be increased. To use the interactive companion and not use the textbook problems as homework would diminish the student's skill levels in Geometry to an unacceptably low level.

Daniel C. Alexander

Statements

1. Complete: A _____ is a group of words and/or symbols that can collectively be classified as true or false.

2. Which of the following is (are) statements? _____.
   (i) 3 + 4 = 12   (ii) 3X4 = 12   (iii) Are you leaving?
   (iv) Look out!   (v) Billy Bob and his brother

3. For simple statements P and Q, write the:
   (i) Conjunction _____   (ii) Disjunction _____.

4. Classify these statements as true or false.
   a. Babe Ruth played baseball and 3 + 4 = 12._____.
   b. Babe Ruth played baseball and 3X4 = 12._____.
   c. Babe Rurh played baseball or 3 + 4 = 12. _____.
   d. Babe Ruth played baseball or 3X4 = 12._____.

5. State the negation of the given statement.
   a. Mary is a seamstress. _____.
   b. Billy Bob is not on the honor roll this semester.
       _____.

6. If statement P is true while statement Q is false, classify these statements as true or false.
   (i) not P _____   (ii) not Q _____.
   (iii) P and not Q _____   (iv) not P or Q _____.

7. Classify each implication (conditional) statement as true or false.
   (i) If x is an even integer, then it can be divided exactly by 2. _____.
   (ii) If Antonio lives in Dallas, then he lives in Texas. _____.

Induction, Intuition, and Deduction

8. In the following statement, what does intuition suggest that you may conclude?
   M is the midpoint of line segment AB.

   A   M   B

   _____.

9. In the following situation, what does induction suggest that you may conclude?
   It rained Sunday, Monday, Tuesday, and Wednesday. What do you expect of Thursday's weather?
   _____.

10. In the following situation, what does deduction allow you to conclude? Assume that statements (i) and (ii) are both true?
    i.   If a person grew up in Minnesota, then he/she has seen snow.
    ii. Katrina was raised in Rochester, Minnesota.

    _____.

11. In the following situation, what does deduction allow you to conclude? Assume that statements (i) and (ii) are both true?
    i.   If a person lives in Acapulco, then he/she lives in Mexico.
    ii. Pablo lives in Mexico

    _____.

12. In the following situation, what does deduction allow you to conclude? Assume that statements (i) and (ii) are both true?
    i.   All professional baketball players are tall.
    ii. MJ is a professional basketball player.

    _____.

Venn Diagrams

13. Let P represent the set of "members of the state house of representatives" while Q represents "legislators with 2 year terms of office." Which Venn diagram represents correctly implication, "If a person is a member of the state house of representatives, then he/she is a legislator with a 2 year term of office."
    The Venn diagram on the _____ correctly represents the situation.

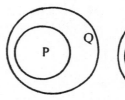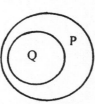

14. Use a Venn diagram to draw a conclusion, if one is possible, for the following situation. Assume that statements (i) and (ii) are true.
    i.   If an object is a skeegle, then it will store water.
    ii. A bonf is a skeegle.

    _____.

15. Use a Venn diagram to draw a conclusion, if one is possible, about the angles of triangle ABC. Assume that statements (i), (ii) and (iii) are true.
    i.   If a figure is an isosceles triangle, then the triangle has 2 sides of the same length.
    ii.   If a triangle has 2 sides of the same length, then it has 2 angles of the same measure.
    iii. Triangle ABC is an isosceles triangle.

    _____.

Undefined Terms

1. Three of the undefined terms of geometry are these building blocks: point, _____, and plane.

2. The symbol A-X-B indicates that points A, X, and B are collinear with point __ between A and B.

Symbols

3. The symbol ∠ represents the word _____ while the symbol △ represents the word _____.

4. While $\overline{CD}$ represents a line segment (a set of points), the length of the line segment is ____.

Measuring Line Segments

5. With $\overline{RS}$ as shown, its length is __ centimeters.

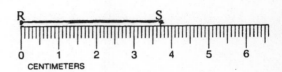

6. In the figure for Exercise 5, at what distance does the midpoint of $\overline{RS}$ lie form R or S?

7. In the figure at the right, suppose that AB = 5.6 and BC = 9.2. Then AC = ___.

8. In the figure at the right, suppose that AC = 15 and BC = 9. Then AB = ___.

Measuring Angles

9. The instrument used to measure an angle is called a _____.

10. On the protractor shown, find the measure of:
    (i) ∠ DBC _____   (ii) ∠ ABD _____.

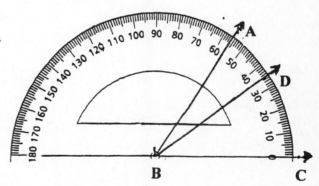

11. In the figure provided, find:
    (i)   m ∠ ABD if m ∠ ABC = 32° and m ∠ CBD = 29°
    (ii)   m ∠ ABC if m ∠ ABD = 62° and m ∠ CBD = 28°
    (iii)   m ∠ ABC if m ∠ ABD = 64° and $\overline{BC}$ bisects ∠ ABD.

12. The sides of a _____ angle are in opposite directions.

13. The sides of a _____ angle are perpendicular.

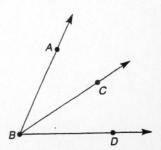

Constructions

14. The instruments used to perform a construction are the straightedge and _____.

15. The compass can be used to draw circles or the part of a circle known as a(n) _____.

16. Construct a line segment that is congruent to $\overline{AB}$.   A _____ B

17. Construct (locate) the midpoint of $\overline{CD}$.
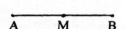
C   D

Further Problems

18. In the figure, M is the midpoint of $\overline{AB}$.
    If AM = 3x − 5 and MB = x + 7, find x.   A   M   B

19. In the figure, RS = x and ST = y.
     If RT = 17 and RS − ST = 3, find x and y.   R   S   T

Mathematical Systems

   1. The four parts of a mathematical system are undefined terms, _____, postulates, and _____.

   2. Some examples of mathematical systems are algebra, _____, and calculus.

A Good Definition

   3. A good definition places the term being defined into a category. In the following definition, what is the category? "An isosceles triangle is a triangle with 2 congruent sides."

   4.   A definition must be reversible. For this reason, the statement "If 2 angles are congruent, then these angles have equal measures" has what counterpart?

Initial Postulates

   5. Complete: Through two distinct points, there is one _____.

   6. Complete: (Ruler Postulate) The measure of any line segment is a unique _____ number.

   7. Complete: Segment-Addition Postulate: If X is on $\overline{AB}$ and A-X-B, then _____.

   8. Complete: If two lines intersect, they intersect in a _____.

Definitions

   9. If $\overline{AB}$ and $\overline{CD}$ are congruent, then _____.

A _____ B

C _____ D

10, Parallel lines are lines that lie in the same plane but do not _____.

Symbols

11. Draw a line segment to match the description with the symbol:

   (i)   line AB $\overrightarrow{AB}$

   (ii)   line segment AB $\overleftrightarrow{AB}$

   (iii)   length of line segment AB   AB

   (iv)   ray AB $\overline{AB}$

12. Write this statement in symbols:
   (i) Line segment RS is congruent to line segment XY. _____.
   (ii) Lines r and s are parallel.

Planes

13. Coplanar points lie on the same _____.

14. In the figure, the horizontal planes R and S are _____.

15. Complete: Through three noncollinear points, there is exactly one _____.

16. Complete: If two planes intersect, they intersect in a _____.

Further Problems

17. By choosing two points (like A and B) at a time, what is the total number of lines determined by noncollinear points A, B, C, and D? _____

18. Points R, S, and T (not shown) are collinear with RS = 12 and ST = 17. What are the two possible lengths of $\overline{RT}$?

Constructions

19. Given $\overline{XY}$, construct a line segment $\overline{XM}$ so that XM = $\frac{1}{2}$ (XY).

X   Y

20. Using $\overline{XY}$ as shown in problem 19, construct a line segment $\overline{XR}$ so that XR = 2(XY).

Angle Postulates

1. Complete (Protractor Postulate): The measure of an angle is a unique _____ number.

2. Complete (Angle-Addition Postulate): If point D lies in the interior of $\angle$ ABC, then m $\angle$ ABD + m $\angle$ DBC = _____.

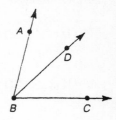

Types of Angles

3. An angle whose measure is less than 90^{0} is a(n) _____ angle.

4. An angle whose measure is 90^{0} is a(n) _____ angle.

5. An angle whose measure is between 90^{0} and 180^{0} is a(n) _____ angle.

6. An angle whose measure is 180^{0} is a(n) _____ angle.

Pairs of Angles

7. Two angles are _____ if their measures are equal.

8. As shown, $\angle$ MNP and $\angle$ PNQ are _____ angles.

9. If $\angle$ MNP $\cong$ $\angle$ PNQ in the figure, then $\overline{NP}$ is said to _____ $\angle$ MNQ.

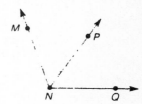

10. Two angles whose sum of measures is 90^{0} are _____ angles.

11. Two angles whose sum of measures is 180^{0} are _____ angles.

12. In the figure, $\angle$ 7 and $\angle$ 8 are _____ angles.

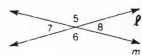

Constructions

13. Construct an angle congruent to $\angle$ RST.

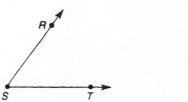

14. Construct the bisector of $\angle$ PRT.

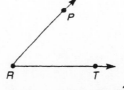

Further Problems

15. If m $\angle$ ABD = 29^0 and m $\angle$ DBC = 42^0, then m $\angle$ ABC = _____.

16. If m $\angle$ ABD = 29^0 and m $\angle$ ABC = 72^0, then m $\angle$ DBC = _____.

17. If m $\angle$ ABD = x, m $\angle$ DBC = 2x - 3, and m $\angle$ ABC = 72, then x = _____.

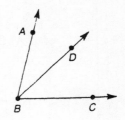

Exs 15-17

18. If m $\angle$ 7 = 27^0, then m $\angle$ 8 = ___ and m $\angle$ 5 = _____.

19. If m $\angle$ 7 = 2x + 9 and m $\angle$ 8 = 5x – 21, then x = ___ .

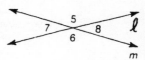

20. If m $\angle$ 7 = 2x - 11 and m $\angle$ 5 = 5(x + 6), then x = ___ .

Exs 18-20

Properties of Equality:

1. Complete each property:
    (i)   Addition Property of Equality: If $a = b$, then $a + c =$ _____.
    (ii) Subtraction Property of Equality: If $a = b$, then _____.
    (iii) Multiplication Property of Equality: If $a = b$, then _____.
    (iv) Division Property of Equality: If $a = b$ and $c \neq 0$, then _____.

2. (i) Use the Addition Property of Equality to draw a conclusion: If $2x - 3 = 7$, then _____.
    (ii) Use the Division Property of Equality to draw a conclusion: If $2x = 10$, then _____.

Further Algebraic Properties

3. Where a, b, and c are real numbers, name each Property of Equality.
    (i) $a(b + c) = ab + ac$ _____.
    (ii) If $a = b$, then a can replace b in any equation. _____.
    (iii) If $a = b$ and $b = c$, then $a = c$. _____.

4. Complete each statement:

    (i) $2(x + 5) =$ _____   (ii) Because $\frac{1}{2} = 0.5$ and $0.5 = 50\%$, then _____.

Algebraic Proof

5. In a proof, we make claims called _____ and support these with reasons that include
   Given, definitions, postulates, and theorems.

6. Complete the REASONS in this algebraic proof:
    Given: $2(x - 5) + 3 = 17$
    Prove: $x = 12$
        Proof

Statements	Reasons
(1) $2(x - 5) + 3 = 17$	(1) _____
(2) $2x - 10 + 3 = 17$	(2) _____
(3) $2x - 7 = 17$	(3) _____
(4) $2x = 24$	(4) _____
(5) $x = 12$	(5) _____

7. Complete the missing STATEMENTS/REASONS in this algebraic proof:
    Given: $2x - 9 = 3x - 15$
    Prove: $x = 6$ (same as $6 = x$)
        Proof

Statements	Reasons
(1) _____	(1) Given
(2) $2x + 6 = 3x$	(2) _____
(3) _____	(3) _____

8. In a geometric proof, the statements are written in an orderly sequence and each statement is
    supported by a reason. Given the drawing at the right,
    for what reason can you say that:

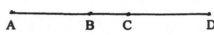

A　　　　　B　C　　　　　D

    (i) AB + BC = AC? _____ .

    (ii) If AB = CD, then AB + BC = BC + CD? _____ .

9. In the drawing for Exercise 8, suppose that AB = 9, BC = 2, and CD = 9. Does AC = BD? _____ .

10. Complete the REASONS in this geometric proof. Refer to the drawing as needed.

    Given: AB = CD on $\overline{AD}$
    Prove: AC = BD

A　　　　　B　C　　　　D

Proof

Statements	Reasons
(1) AB = CD on $\overline{AD}$	(1) _____ .
(2) AB + BC = BC + CD	(2) _____ .
(3) AB + BC = AC and BC + CD = BD	(3) _____ .
(4) AC = BD	(4) _____ .

11. Consider the drawing in the following exercise. Suppose that m $\angle$ RSW = 43^0, m $\angle$ VSW = 21^0,
    and m $\angle$ VST = 43^0. Does m $\angle$ RSV = m $\angle$ WST? _____ .

12. Complete the missing STATEMENTS/REASONS in this geometric proof.

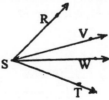

    Given: m $\angle$ RSW = m $\angle$ VST
    Prove: m $\angle$ RSV = m $\angle$ WST

Proof

Statements	Reasons
(1) _____	(1) _____ .
(2) m $\angle$ RSW = m $\angle$ RSV + m $\angle$ VSW and m $\angle$ VST = m $\angle$ WST + m $\angle$ VSW	(2) _____ .
(3) m $\angle$ RSV + m $\angle$ VSW = m $\angle$ WST + $\angle$ VSW	(3) Substitution Prop. of Equality
(4) _____ .	(4) _____ .

Perpendicular Lines

1. How are a vertical line m and a horizontal line n related? _____.

2. If two lines are perpendicular, they form _____ angles.

Relations

3. The relationship "is perpendicular to" relates line while the relation "is equal to" relates _____.

4. The three special properties used to characterize a relation R are:
   (i) _____ Property: a R a
   (ii) _____ Property: If a R b, then b R a.
   (iii) _____ Property: If a R b and b R c, the a R c.

5. The Transitive Property of a given relation, when one exists, can be used to relate more than three objects. In the sequence of statements, the first object named can be related to the last object named. For example, if $\angle 1 \cong \angle 2$, $\angle 2 \cong \angle 3$, and $\angle 3 \cong \angle 4$, then _____.

6. For the relation "is congruent to" when used to compare angles, which properties (Reflexive, Symmetric, Transitive) exist? _____.

7. For the relation "is greater than" when used to compare numbers, which properties (Reflexive, Symmetric, Transitive) exist? _____.

8. For the relation "is perpendicular to" when used to compare lines, which properties (Reflexive, Symmetric, Transitive) exist? _____.

9. For the relation "is equal to" when used to compare numbers, which properties (Reflexive, Symmetric, Transitive) exist? _____.

Constructions

10. (i) Given point P on line t in plane Q, how many lines can be drawn in plane Q that perpendicular to t at point P? _____.
    (ii) Given point P on line t, how many lines can be drawn in space that are perpendicular to t at point P? _____.

11. Construct the line perpendicular to $\overleftrightarrow{AB}$ at point P.

12. Construct the perpendicular-bisector of $\overline{AB}$.

A •————————• B

Further Problems

13. Supply/complete missing REASONS for this proof problem. This problem verifies the theorem, "If two lines are perpendicular, they meet to form a right angle."

Given: $\overline{AB} \perp \overline{CD}$, intersecting at point E
Prove: $\angle AEC$ is a right angle.

Proof

Statements	Reasons
(1) $\overline{AB} \perp \overline{CD}$, intersecting at point E	(1) _____.
(2) $\angle AEC \cong \angle CEB$	(2) _____ lines form $\cong$ adjacent angles.
(3) $m \angle AEC = m \angle CEB$	(3) _____.
(4) $\angle AEB$ is a straight angle, so $m \angle AEB = 180^0$	(4) _____.
(5) $m \angle AEC + m \angle CEB = m \angle AEB$	(5) _____.
(6) $m \angle AEC + m \angle CEB = 180^0$	(6) _____.
(7) $m \angle AEC + m \angle AEC = 180^0$ or $2 \cdot m \angle AEC = 180^0$	(7) _____.
(8) $m \angle AEC = 90^0$	(8) _____.
(9) $\angle AEC$ is a right angle.	(9) _____.

14. In Example 3 of Section 1.6, we verify this theorem:
   "If two lines intersect, then the vertical angles formed are congruent."
   Use this theorem to solve each problem.

   (i)   If $m \angle 1 = 142^0$, then $m \angle 3 =$ _____.
   (ii) If $m \angle 1 = x$, then $m \angle 3 =$ _____.
   (iii) If $m \angle 1 = x$, then $m \angle 2 =$ _____.
   (iv) If $m \angle 2 = 2x + 3$ and $m \angle 4 = 5x - 33$, then $x =$ _____.

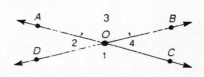

Hypothesis and Conclusion

1. In the conditional statement, "If P, then Q," The hypothesis is simple statement ___ and the conclusion is the simple statement ___.

2. In the following statement, underline the hypothesis once and the conclusion twice.
"If two lines intersect, the vertical angles formed are congruent."

3. Rewrite the following statement in the form of a conditional statement.
"All isosceles triangles have a pair of congruent sides."

_____.

The Written Parts of a Formal Proof

4.   The five parts that must be shown in the formal proof of a theorem are:
(i) Statement of the _____,
(ii) Drawing representing the facts in the _____ of the theorem,
(iii) Given, which describes the drawing based upon the _____ of the theorem,
(iv) Prove, which describes the drawing based upon the _____ of the theorem, and
(v) Proof, which provides statements and reasons in a logical order – beginning with the
_____ statement and ending with the _____ statement.

5. For the stated theorem and for the drawing provided, write the Given and Prove.

"If two lines meet to form a right angle, then these lines are perpendicular."

Given: _____.
Prove: _____.

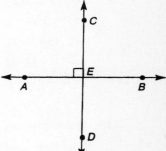

Converse of a Statement

6. The converse of the statement, "If P, then Q" is the statement _____

7. Is the converse of every conditional statement "If P, then Q" necessarily true? _____.

8. (i) Write the converse of the statement, "If two lines are perpendicular, they meet to form a right angle." _____.
(ii) Is the converse that you wrote in part (i) true or false. _____.

Theorems of Section 1.7

9. Suppose that (i) $\angle 1$ is complementary to $\angle 3$ and that
(ii) $\angle 2$ is complementary to $\angle 3$.
How are $\angle 1$ and $\angle 2$ related? _____.

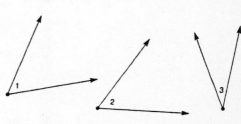

10. Consider the theorem, "If the exterior sides of two adjacent acute angles form perpendicular rays, then these angles are complementary."
In the drawing, $\overline{BA} \perp \overline{BC}$. If it is known that m$\angle 1 = 28^0$, use this theorem to find m$\angle 2$. _____.

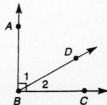

11. Complete the missing statements and reasons in the formal proof of the following theorem.
"If two angles are supplementary to the same angle, then these angles are congruent."

Given: $\angle 1$ is supp. to $\angle 2$
$\angle 3$ is supp. to $\angle 2$
Prove: $\angle 1 \cong \angle 3$

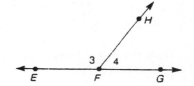

Proof

Statements	Reasons
(1) $\angle 1$ is supp. to $\angle 2$ $\angle 3$ is supp. to $\angle 2$	(1) _____.
(2) $m\angle 1 + m\angle 2 = 180$ $m\angle 3 + m\angle 2 = 180$	(2) _____.
(3) $m\angle 1 + m\angle 2 = $ _____.	(3) Substitution Prop. of Equality
(4) $m\angle 1 = m\angle 3$	(4) _____ Prop. of Eq.
(5) _____.	(5) _____.

12. Consider the theorem, "If the exterior sides of two adjacent angles form a straight line, then these angles are supplementary."

In the drawing, $\overline{EG}$ is a straight line.

(i) If it is known that $m\angle 3 = 128^0$, find $m\angle 4$. _____.

(ii) If it is known that $m\angle 4 = 49^0$, find $m\angle 3$. _____.

(iii) If $m\angle 3 = 3y$ and $m\angle 4 = y$, find y. _____.

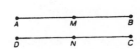

13. In the drawing, $\overline{AB} \cong \overline{DC}$, M is the midpoint of $\overline{AB}$, and
N is the midpoint of $\overline{DC}$. How are the four line segments $\overline{AM}$,
$\overline{MB}$, $\overline{DN}$, and $\overline{NC}$ related? _____.

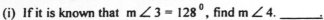

14. In the drawing, $\angle ABC \cong \angle EFG$. Also, $\overline{BD}$ bisects $\angle ABC$
and $\overline{FH}$ bisects $\angle EFG$. If $m\angle ABC = 56^0$, find the measure
of each of the numbered angles. _____.

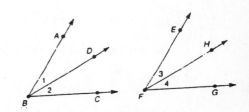

Parallel and Perpendicular Lines

1. In the drawing, how many lines can be drawn from point P
    that are perpendicular to line ℓ ? _____ .

2. In the drawing, how many lines can be drawn from point P
    that are parallel to line ℓ ? _____ .

3. In a plane, line m is drawn perpendicular to line t. In that plane,
    a second line n is drawn perpendicular to line t. How are lines
    m and n related? _____ .

Lines with a Transversal

4. In the drawing, use numbers to name four angles
    that are interior angles. _____ .

5. In the drawing, use numbers to name two angles that
    are corresponding angles. _____ .

6. In the drawing, use numbers to name two angles that
    are alternate interior angles. _____ .

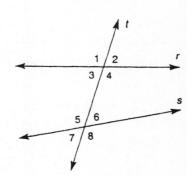

Parallel Lines and Angles

7. Complete this postulate: If two _____ lines are cut by a transversal,
    then the corresponding angles are congruent.

8. Use the theorem, "If two parallel lines are cut by a transversal,
    then the alternate interior angles are congruent" to determine
    which angle is congruent to $\angle 3$. _____ .

9. Complete this theorem: If two parallel lines are cut by a transversal, then the
    exterior angles on the same side of the transversal are _____ .

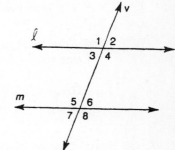

10. Fill in missing statements and reasons for the proof of this theorem:
    "If two parallel lines are cut by a transversal, then the alternate interior angles are congruent."

    Given: a ∥ b; transversal k
    Prove: $\angle 3 \cong \angle 6$

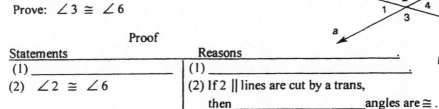

Proof	
Statements	Reasons
(1) _____	(1) _____ .
(2) $\angle 2 \cong \angle 6$	(2) If 2 ∥ lines are cut by a trans,
	then _____ angles are $\cong$.
(3) $\angle 3 \cong \angle 2$	(3) If 2 lines intersect, the _____ angles
	formed are congruent.
(4) _____ .	(4) Transitive Prop. of $\cong$ (for angles)

Further Problems

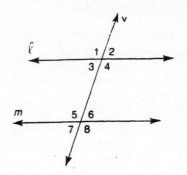

11. In the figure, $\ell \parallel m$ and transversal v. Find:

    (i) m $\angle$ 5 if m $\angle$ 1 = 118^0. _____.

    (ii) m $\angle$ 5 if m $\angle$ 8 = 115^0. _____.

    (iii) m $\angle$ 2 if m $\angle$ 8 = 117^0. _____.

    (iv) m $\angle$ 1 if m $\angle$ 3 = 3(m $\angle$ 1). _____.

12. In the figure, a $\parallel$ b and transversal k. Find:

    (i) x, if m $\angle$ 5 = x and m $\angle$ 1 = 42^0. _____.

    (ii) y, if m $\angle$ 5 = 3x + 13 and m $\angle$ 8 = 4x + 3. _____

    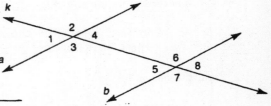

    (iii) z, if m $\angle$ 2 = 5(z + 3) and m $\angle$ 8 = 7(z - 5). _____

    (iv) w, if m $\angle$ 3 = 7w – 10 and m $\angle$ 5 = 70 – w. _____

13. For this problem, assume that any postulate or theorem of Section 2.1 can be cited as a reason. Complete the missing statements and reasons of this proof problem.

    Given: $\ell \parallel m$ and m $\parallel$ n
    Prove: $\angle$ 1 $\cong$ $\angle$ 4

    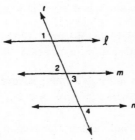

	Proof	
Statements	Reasons	
(1) $\ell \parallel m$	(1) _____.	
(2) $\angle$ 1 $\cong$ $\angle$ 2	(2) If 2 $\parallel$ lines are cut by a trans, the _____. $\angle$ a are $\cong$.	
(3) $\angle$ 2 $\cong$ $\angle$ 3	(3) _____.	
(4) _____	(4) Given	
(5) $\angle$ 3 $\cong$ $\angle$ 4	(5) _____.	
(6) _____	(6) Transitive Prop. of $\cong$ (for angles)	

Statements Written in Symbolic Form

1. Given that the conditional statement "If P, then Q" is written in symbolic form
    as P $\rightarrow$ Q while "not P" is written ~ P, write each statement in symbolic form:
    (i) Converse of "If P, then Q" _____.
    (ii) Inverse of "If P, then Q" _____.
    (iii) Contrapositive of "If P, then Q" _____.

2. Given that conditional statement, "If P, then Q" is true, which statement (converse, inverse, or
    Contrapositive) *must* also be true? _____.

Law of Negative Inference (Law of Contraposition)

3. Complete the conclusion in the Law of Negative Inference:
    (1) P $\rightarrow$ Q
    (2) ~ Q
    (C) _____.

4. Use the Law of Negative Inference to state the conclusion in this argument.
    (1) If Pablo lives in Guadalajara, then he lives in Mexico.
    (2) Pablo does not live in Mexico.
    (C) _____.

When to use Indirect Proof

5. Indirect proof is often used when statement Q of the theorem "If P, then Q" has the form
    of a _____.

6. For which of the following "Prove" statements would you be likely to use an indirect proof?
    (i) c $\neq$ d   (ii) c = d   (iii) $\angle$ s 1 and 2 are not congruent.

    _____

7. Indirect proof is also used to prove "uniqueness" theorems. For which theorems would you use an
    indirect proof? _____.
    | | |
    | (i) | The midpoint of a line segment is unique. |
    | (ii) | An angle has exactly one angle-bisector. |
    | (iii) | Through a point outside a line, there is only one line parallel to the given line. |
    | (iv) | All right angles are congruent. |

The Method of Indirect Proof

8. To use indirect proof with a problem of the form:
    Given: P
    Prove: Q,
    we use these steps:
    (i) Suppose that _____ is true.
    (ii) Reason from the supposition in statement (i) until you reach a _____.
    (iii) Note that the supposition must be false and that statement _____ must therefore be true.

9. Fill blanks to complete the proof of the following statement. No drawing is provided!
    "If $\angle 1 \not\cong \angle 2$, then $\angle 1$ and $\angle 2$ are not vertical angles."
        Given: $\angle 1 \not\cong \angle 2$
        Prove: $\angle 1$ and $\angle 2$ are not vertical angles."

    Proof: Suppose that $\angle 1$ and $\angle 2$ are _____.
    Then _____ because vertical angles are congruent.
    However, it is given that $\angle 1 \not\cong \angle 2$, which contradicts the congruence
    of vertical angles. The supposition must be _____, so it
    follows that _____.

10. Fill blanks to complete the proof of the following statement.

    "The bisector of an angle is unique."
        Given: $\overline{BD}$ bisects $\angle ABC$
        Prove: $\overline{BD}$ is the only angle-bisector for $\angle ABC$

    Proof: $\overline{BD}$ bisects $\angle ABC$, so m $\angle ABD = \dfrac{1}{2}$ _____.

    Suppose that $\overline{BE}$ also bisects $\angle ABC$, so m $\angle ABE = \dfrac{1}{2}$ m $\angle ABC$.

    By the _____ Postulate, m $\angle ABD =$ m $\angle ABE +$ m_____.

    By the reason _____, the preceding statement becomes

    $\dfrac{1}{2}$ m $\angle ABC = \dfrac{1}{2}$ m $\angle ABC +$ m $\angle EBD$.

    Applying the Subtraction Property of Equality, m $\angle EBD =$ _____.
    But this contradicts the Protractor Postulate, which states that the measure of any angle is
    a unique _____ number. Due to this contradiction, the supposition must
    be false and it follows that _____.

A Contrast

1. In Section 2.1, the postulates and theorems generally have the form "If two parallel lines are cut by a transversal, then . . . ." Those statements could be used to justify that a pair of angles are _____ or that a pair of angles are _____.

2. In this section (Section 2.3), the theorems take the form "If . . . . , then the two lines are parallel." These statements can be used to justify that _____.

Lines with a Transversal

3. Complete this theorem: If two lines are cut by a transversal so that corresponding angles are congruent, then these lines are _____.

4. Complete this theorem: If two lines are cut by a transversal so that interior angles on the same side of the transversal are _____, then these lines are parallel.

5. Complete this theorem: If two lines are each parallel to a third line, then these lines are _____.

6. Complete this theorem: If two _____ lines are each perpendicular to a third line, then these lines are parallel.

Applications

7. If m $\angle$ 1 = 107^0, find m $\angle$ 5 so that ℓ || m. _____.

8. If m $\angle$ 4 = 106^0, find m $\angle$ 6 so that ℓ || m. _____.

9. If m $\angle$ 2 = 72 and m $\angle$ 7 = 4x + 20, find x so that ℓ || m. _____.

10. If m $\angle$ 3 = 2x + 26 and m $\angle$ 5 = 6(x – 1), find x so that ℓ || m. _____.

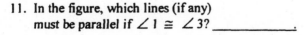

11. In the figure, which lines (if any) must be parallel if $\angle$ 1 $\cong$ $\angle$ 3? _____.

12. In the figure, which lines (if any) must be parallel if $\angle$ 1 $\cong$ $\angle$ 4? _____.

13. In the figure, which lines (if any) must be parallel if $\angle$ 8 $\cong$ $\angle$ 3? _____.

14. In the figure, which lines (if any) must be parallel if $\angle$ 3 and $\angle$ 7 are supplementary? _____.

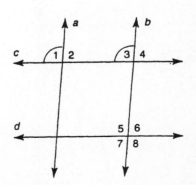

Proving Lines Parallel

15. Fill in missing statements and reasons for the proof of this theorem:
   "If two lines are cut by a transversal so that alternate interior angles
   are congruent, then these lines are parallel.'

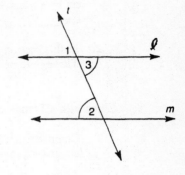

   Given: Lines ℓ and m; transversal t; $\angle 2 \cong \angle 3$
   Prove: $\ell \parallel m$

Proof

Statements	Reasons
1. Lines ℓ and m; trans. t; $\angle 2 \cong \angle 3$	1. _____.
2. $\angle 1 \cong \angle 3$	2. _____.
3. _____	3. Trans. Prop. of Congruence
4. _____	4. If 2 lines are cut by a trans. so that corr. $\angle$ s are $\cong$, these lines are parallel.

16. Fill in missing statements and reasons for this proof problem:

   Given: $\ell \parallel m$; $\angle 3 \cong \angle 4$
   Prove: $\ell \parallel n$

Proof

Statements	Reasons
1. $\ell \parallel m$	1. _____.
2. $\angle 1 \cong \angle 2$	2. _____.
3. $\angle 2 \cong \angle 3$	3. _____.
4. _____	4. Given
5. $\angle 1 \cong \angle 4$	5. Trans. Prop. of Congruence
6. _____	6. If 2 lines are cut by a trans. so that corr. $\angle$ s are $\cong$, these lines are parallel.

A Construction

17. Construct the line parallel to $\overline{AB}$
   that passes through point P.

   • P

Triangle Types Classified by Sides

1. If no sides of a triangle are congruent, the triangle is a(n) _____ triangle.

2. If two sides of a triangle are congruent, the triangle is a(n) _____ triangle

3. If all three sides of a triangle are congruent, the triangle is a(n) _____ triangle.

Triangle Types Classified by Angles

4. If all angles of a triangle are acute, the triangle is a(n) _____ triangle.

5. If one angle of a triangle is a right angle, the triangle is a(n) _____ triangle

6. If one angle of a triangle is obtuse, the triangle is a(n) _____ triangle.

7. If all three angles of a triangle are congruent, the triangle is a(n) _____ triangle.

The Sum of Angles of a Triangle

8. In the figure provided, m $\angle$ 1 + m $\angle$ 2 + m $\angle$ 3 = _____ degrees.

9. In the figure provided, $\overline{ED} \parallel \overline{AB}$. Then $\angle$ 1 $\cong$ $\angle$ A and $\angle$ 3 $\cong$ $\angle$ B because these are pairs of _____
    _____ angles .

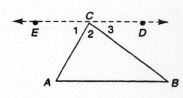

10. The sum of the measures of the interior angles of $\triangle$ ABC
    (or any other triangle) is _____ degrees.

Corollaries of the theorem,"The sum of the measures of the interior angles of a triangle is 180^0 ."

11. Complete: Each angle of an equiangular triangle measures _____ degrees.

12. Complete: The acute angles of a right triangle are _____.

13. Complete: If two angles of one triangle are congruent to two angles of a second triangle,
    then the _____ angles are also congruent.

14. Complete: The measure of an _____ angle of a triangle equals the sum of measures
    of the two nonadjacent interior angles.

Problems

15. In the figure provided, find m $\angle$ 1 + m $\angle$ 2. _____.

16. In the figure provided, find x if m $\angle$ 1 = 4x + 7
    and m $\angle$ 2 = 2x + 3. _____.

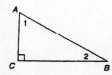

17. In the figure provided, find m $\angle$ 4 if m $\angle$ 2 = 113^0 and

    m $\angle$ 5 = 26^0. _____ .

18. Find an expression for m $\angle$ 1 if m $\angle$ 3 = x and m $\angle$ 5 = y. _____ .

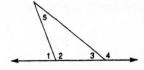

A Sample Proof

19. The following proof is that of the corollary:

    "The measure of an exterior angle of a triangle equals the sum
    of measures of the two nonadjacent interior angles."

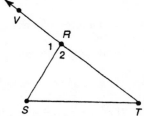

    Given: $\triangle$ RST with exterior angle 1
    Prove: m $\angle$ 1 = m $\angle$ S + m $\angle$ T

                          Proof

Statements	Reasons
1. $\triangle$ RST with exterior angle 1	1. _____ .
2. _____ .	2. If the exterior sides of 2 adjacent $\angle$ s form a straight line, the $\angle$ s are supplementary.
3. m $\angle$ 1 + m $\angle$ 2 = 180^0	3. _____ .
4. m $\angle$ 2 + m $\angle$ S + m $\angle$ T = 180^0	4. _____ .
5._____ .	5. Substitution Prop. of Equality
6._____ .	6. Subtraction Prop. of Equality

Polygons

1. Give the name of the polygon that has:
     3 sides _____ .   4 sides _____ .
     5 sides _____ .   6 sides _____ .
     8 sides _____ .   10 sides _____ .

2. What term is used to describe a polygon that has:
     (i) all sides congruent? _____ . See figure (a)
     (ii) all angles congruent? _____ . See figure (b)
     (iii) all sides congruent and all
          angles congruent? _____ . See figure (c)

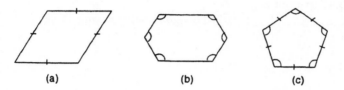

(a)   (b)   (c)

Diagonals of a Polygon

3. Quadrilateral ABCD (not shown) has two diagonals. Name the diagonals. _____ .

4. By drawing and counting, how many diagonals does a pentagon have? _____ .

5. Where n is the number of sides in a polygon, the number of diagonals is D = $\dfrac{n(n-3)}{2}$.

     Find the total number of diagonals for an octagon. _____ .

Sum of the Interior Angles of a Polygon

6. For a polygon with n sides, diagonals can be drawn from a selected vertex to separate the polygon into _____ triangles.

7. The sum of measures of the interior angles of a polygon with n sides is S = _____ .

8. Find the sum of the interior angles of any quadrilateral. _____ .

9. Find the sum of the interior angles of a polygon with 12 sides. _____ .

Regular and Equiangular Polygons

10. Because the sum of measures of all the interior angles of a polygon with n sides
     is $(n-2) \bullet 180^{0}$, the formula for the measure of each interior angle of a regular
     or equiangular polygon is I = _____ .

11. Find the measure of each interior angle of a:
    (i) regular pentagon _____. (ii) equiangular polygon with 12 sides. _____.

12. For a regular octagon, find:
    (i) the measure of each interior angle. _____.
    (ii) the measure of each exterior angle. _____.

Exterior Angles of a Polygon

13. The sum of the exterior angles (one at each vertex) of a polygon of n sides is always _____.

14. Find the measure of each exterior angle of a:
    (i) regular decagon _____. (ii) equiangular polygon with 18 sides. _____.

Polygrams

15. Roughly speaking, a _____ is a star-shaped figure with 5 or more points.

16. Being specific as to type, name each polygram below:
    (i) _____. (ii) _____. (iii) _____.

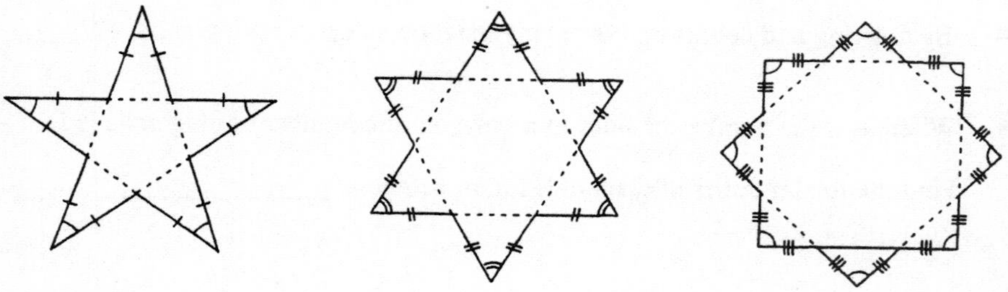

Line Symmetry

1. When a figure has a horizontal symmetry, there is a vertical axis (line) of symmetry. Similarly, a figure that has vertical symmetry has a _____ axis (line) of symmetry.

2. Of the geometry figures shown below, which figures have line symmetry? _____.
   (a) Square   (b) Letter A   (c) Parallelogram   (d) Letter S

A   S

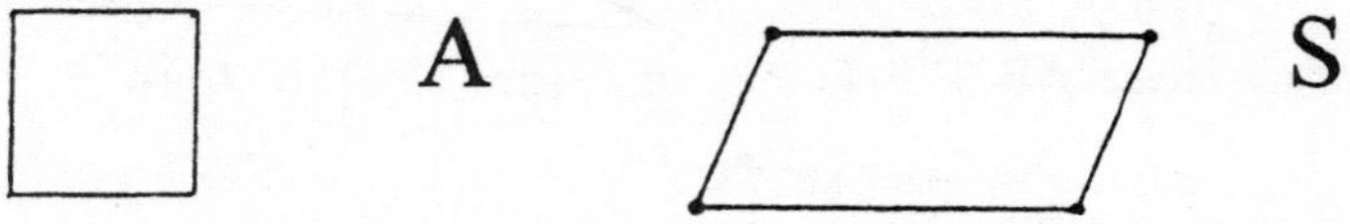

3. Of the figures shown in Exercise 2, do any of these have more than one line of symmetry? _____.

4. Each of the figures shown has line symmetry. Is the line of symmetry vertical, horizontal, both, or neither? (a)_____ ; (b)_____ ; (c)_____ ; (d)_____ .

   (a) Isosceles   (b) Letter E   (c) Rectangle   (d) Letter T
       Trapezoid

E   T

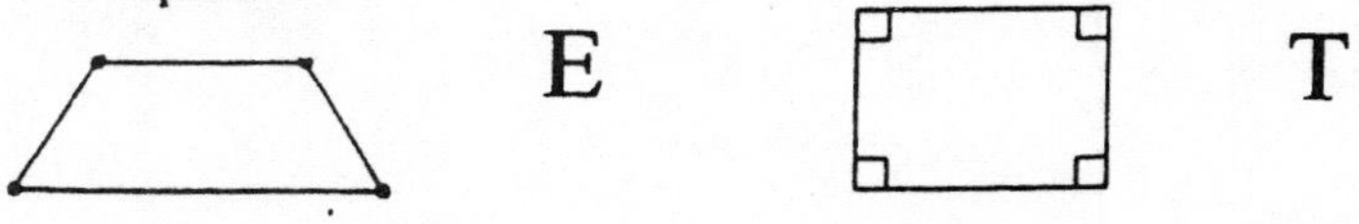

Point Symmetry

5. When a figure has point symmetry with respect to point P, there there is a point C on the figure that corresponds to point A on the figure in such a way that P is the _____ of $\overline{AC}$.

6. Of the geometry figures shown below, which figures have point symmetry? _____.
   (a) Square   (b) Letter A   (c) Parallelogram   (d) Letter S

A   S

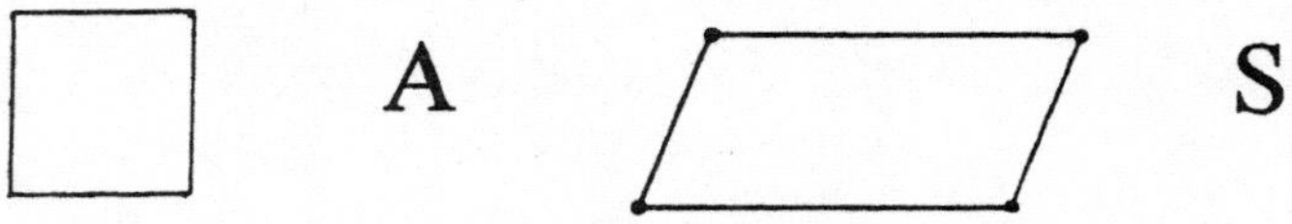

7. Do all regular polygons have point symmetry? _____.

8. Of the figures shown, which have point symmetry? _____.

   (a) Isosceles   (b) Letter E   (c) Rectangle   (d) Letter T
       Trapezoid

E   T

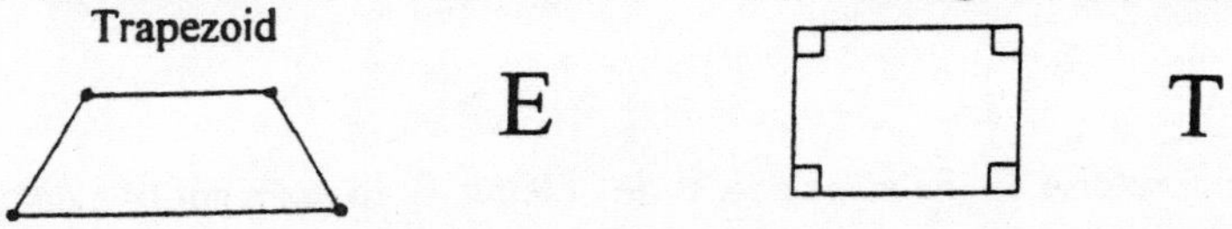

Transformations (Slides)

9. When we slide △ABC to form △DEF, we know that these triangles are _____ .

10. Complete the slide of quadrilateral ABCD to form quadrilateral EFGH if A ↔ E.

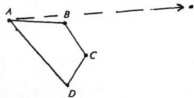

Transformations (Reflections)

11. Complete the reflection of △ABC across the vertical line ℓ .

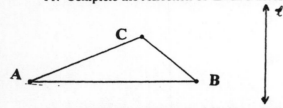

12. Complete the reflection of the letter P across the horizontal line m.

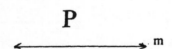

Transformations

13. Complete the rotation of the letter W
    through a 180^0 angle about point P.

W   •P

14. Is the effect of the rotation in Exercise 13 the same as a slide of letter W to the right of a distance
    that is equal to that which is twice the distance from W to P? _____ .

Congruent Triangles

1. Given the statement $\triangle ABC \cong \triangle DEF$, we know that:
    (i) $\angle A$ of $\triangle ABC$ corresponds to _____ of $\triangle DEF$.
    (ii) Vertex C of $\triangle ABC$ corresponds to _____ of $\triangle DEF$.
    (iii) Side $\overline{AC}$ of $\triangle ABC$ corresponds to _____ of $\triangle DEF$.

2. Complete these properties for "congruence of triangles:"
    (i) Reflexive: $\triangle ABC \cong$ _____ .
    (ii) Symmetric: If $\triangle ABC \cong \triangle DEF$, then _____ .
    (iii) Transitive: If $\triangle ABC \cong \triangle DEF$ and $\triangle DEF \cong \triangle GHK$, then _____.

Included Sides and Angles

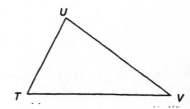

3. In $\triangle UTV$, which:
    (a) side is included by $\angle U$ and $\angle V$? _____ .
    (b) angle is included by $\overline{UV}$ and $\overline{UT}$? _____ .

Reflexive Property of Congruence

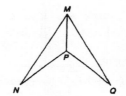

4. In a proof, the statement $\angle A \cong \angle A$ can be justified by, writing the one-word reason _____ .

5. In the figure, what statement can you make that is supported by the reason Identity? _____ .

6. In the figure, what statement can you make that is supported by the reason Identity? _____ .

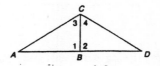

Methods Used to Prove that two Triangles are Congruent

7. The statement, "If the three sides of one triangle are congruent to the three sides of a second triangle, then the triangles are congruent" is indicated by the reason _____.

8. The statement, "If two sides and the included _____ of one triangle are congruent to two sides and the included angle of a second triangle, then the triangles are congruent" is indicated by the reason SAS.

9. The four methods for proving a pair of triangles congruent are SSS, SAS, _____, and _____.

10. The figure at the right illustrates why there is no method of proving triangles congruent that is abbreviated _____.

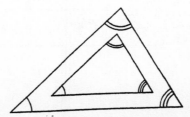

11. The figure at the right illustrates why
    there is no method of proving triangles
    congruent that is abbreviated _____.

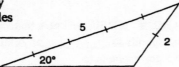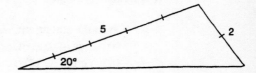

Proving Triangles Congruent

12. With congruent parts of triangles indicated, what method (SSS, SAS, ASA, or AAS)
    would be used to show that the triangles are congruent?
    (i) _____.   (ii) _____.

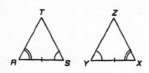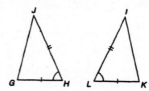

13. Provide missing statements and reasons for the following proof problem.

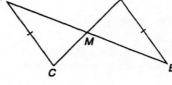

Given: $\overline{AB}$ and $\overline{CD}$ bisect each other at M;

      also, $\overline{AC} \cong \overline{DB}$

Prove: $\triangle AMC \cong \triangle BMD$

Proof

Statements	Reasons
1. $\overline{AB}$ and $\overline{CD}$ bisect each other at M	1. _____.
2. $\overline{AM} \cong \overline{MB}$ and $\overline{CM} \cong \overline{MD}$	2. _____.
3. $\overline{AC} \cong \overline{DB}$	3. _____.
4. _____.	4. _____.

14. Provide missing statements and reasons for the following proof problem.

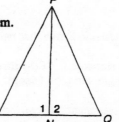

Given: $\overline{PN} \perp \overline{MQ}$; $\overline{MN} \cong \overline{NQ}$

Prove: $\triangle PNM \cong \triangle PNQ$

Proof

Statements	Reasons
1. $\overline{PN} \perp \overline{MQ}$	1. _____.
2. $\angle 1 \cong$ _____.	2. If 2 lines are $\perp$, they form $\cong$ adj. $\angle$ s.
3. _____.	3. Given
4. $\overline{PN} \cong \overline{PN}$	4. _____.
5. _____.	5. _____.

Corresponding Parts of Congruent Triangles

1. From the definition of congruent triangles, the statement "Corresponding parts of congruent triangles are congruent" is expressed in abbreviated form by _____.

2. In the figure, $\triangle ABC \cong \triangle DEF$ by the reason SAS.
    By CPCTC, we know that $\overline{AC} \cong$ _____.

3. In the figure, $\triangle ABC \cong \triangle DEF$ by the reason SAS.
    By CPCTC, we know that $\angle A \cong$ _____.

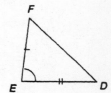

Proof: Using CPCTC

4. If CPCTC is used as a reason in a proof to show that a pair of angles (or sides) of two triangles are congruent, one must first show that the two triangles are _____.

5. Complete missing statements and reasons in the following proof problem.

    Given: $\overline{WZ}$ bisects $\angle$ TWV; $\overline{WT} \cong \overline{WV}$
    Prove: $\overline{TZ} \cong \overline{VZ}$

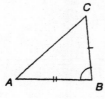

Proof

Statements	Reasons
1. $\overline{WZ}$ bisects $\angle$ TWV	1. _____.
2. _____.	2. The bisector of an angle separates the angle into 2 congruent angles.
3. $\overline{WT} \cong \overline{WV}$	3. _____.
4. _____.	4. Identity
5. $\triangle TWZ \cong \triangle VWZ$	5. _____.
6. _____.	6. CPCTC

6. If the preceding proof problem had asked that we "Prove: Z is the midpoint of $\overline{TV}$,"
    We would include a seventh step. Complete this step.

Statements	Reasons
7. _____.	7. _____.

Suggestions for Proving Triangles Congruent

7. Some suggestions for proving triangles congruent include marking corresponding parts in a like manner. Mark the figures systematically, using:
    a) a _____ in the opening of a right angle;
    b) the same number of dashes on _____ sides; and
    c) the same number of arcs on congruent _____.

8. Consider the markings in the figure at the right.
    a) What type of angle is $\angle$ WZX? _____.
    b) Which side of $\triangle$ ZYX is congruent to $\overline{WZ}$ of $\triangle$ XWZ? _____.

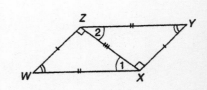

9.   Complete missing statements and reasons in the following proof problem.

Given: $\overline{ZW} \cong \overline{YX}$; $\overline{ZY} \cong \overline{WX}$
Prove: $\overline{ZY} \parallel \overline{WX}$

Proof

Statements	Reasons
1. $\overline{ZW} \cong \overline{YX}$; $\overline{ZY} \cong \overline{WX}$	1. _____
2. _____.	2. Identity
3. _____.	3. SSS
4. $\angle 1 \cong \angle 2$	4. _____.
5. _____.	5. If 2 lines are cut by a trans. so that alt. int $\angle$s are $\cong$, then these lines are parallel

Right Triangles

10. In a right triangle, the sides that form the right angle are called the _____ of the right triangle. The remaining side lies opposite the right angle and is the _____ of the right triangle.

11. The method HL can only be used to prove that two _____ triangles are congruent.

12. Pythagorean Theorem: The square of the length c of the hypotenuse is equal to the sum of squares of the lengths (a and b) of the two legs of the right triangle; that is, $c^2 =$ _____.

13. Square Roots Property: Let x represent the length of a line segment and let p represent some positive number. If $x^2 = p$, then x = _____.

14. In the right triangle shown:
    a) Find c if a = 3 and b = 4. _____.
    b) Find b if c = 10 and a = 8. _____.
    c) Find c if a = 4 and b = 5. _____.
    d) Find a if c = 12 and b = 7. _____.
    Note: In parts c) and d), leave answers in square root form.

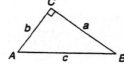

Segments, Rays, Lines Related to the Triangle

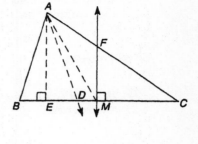

1. If M is the midpoint of side $\overline{BC}$ of △ABC, then
    $\overline{AM}$ is the _____ from vertex A of △ABC.

2. If M is the midpoint of side $\overline{BC}$ of △ABC, then
    $\overline{FM}$ is the perpendicular-bisector of side ____ of △ABC.

3. If ∠BAD ≅ ∠DAC, then $\overline{AD}$ is the angle-bisector
    of _____ of △ABC.

4. If $\overline{AE}$ ⊥ side $\overline{BC}$ of △ABC, then $\overline{AE}$ is the
    _____ from vertex A of △ABC.

5. Every triangle has three angle-bisectors, _____ medians, _____ altitudes,
    and three perpendicular-bisectors of the _____ of the triangle.

6. Complete: Corresponding altitudes of congruent triangles are _____.

Isosceles Triangles

7. In an isosceles triangle, the two sides of equal length are the _____ of the triangle and the
    remaining side is called the _____ of the triangle.

8. In △MNP, $\overline{MP}$ ≅ $\overline{PN}$. Which side of isosceles
    △ MNP is its base? _____.

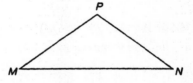

9. In △MNP, $\overline{MP}$ ≅ $\overline{PN}$. Which angle of isosceles
    △ MNP is its vertex angle? _____.

10. Complete missing statements and reasons in the proof of this theorem:
    "The bisector of the vertex angle of an isosceles triangle separates the triangle
    into two congruent triangles."

    Given: △ABC with $\overline{AB}$ ≅ $\overline{BC}$;
           $\overline{BD}$ bisects ∠ABC
    Prove: △ABD ≅ △CBD

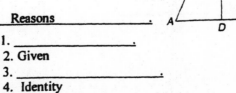

Proof

Statements	Reasons
1. △ABC with $\overline{AB}$ ≅ $\overline{BC}$	1. _____.
2. _____.	2. Given
3. ∠1 ≅ ∠2	3. _____.
4. _____.	4. Identity
5. _____.	5. _____.

11. Complete this theorem: If two sides of a triangle are congruent,
    then the angles that lie _____ these sides are congruent.

12. Complete this theorem: If two angles of a triangle are congruent,
    then the sides that lie opposite these _____ are congruent.

Select Problems

13. In △TUV, $\overline{TV} \cong \overline{UV}$. Which angles of △TUV
    are congruent? _____.

14. In △TUV, ∠T ≅ ∠U. Which sides of △TUV
    are congruent? _____.

15. In △TUV, $\overline{TV} \cong \overline{UV}$. If m∠T = 68°, find:
    a) m∠U. _____. b) m∠V. _____.

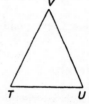

16. In △TUV, $\overline{TV} \cong \overline{UV}$. If m∠V = 54°, find:
    a) m∠T. _____. b) m∠V. _____.

17. In △TUV, $\overline{TV} \cong \overline{UV}$. If m∠T = x + 12 and m∠V = x,   Exercises 13-19
    Find: a) x   _____.   b) m∠T _____.

18. In △TUV, $\overline{TV} \cong \overline{UV}$. If TV = 12 and TU = 9,
    find the perimeter of △TUV. _____.

19. In △TUV, $\overline{TV} \cong \overline{UV}$. If TV = x and TV = 2x - 3, and
    the perimeter of △TUV is 39, find x. _____.

Corollaries

20. Complete: An equilateral triangle is also _____.

21. Complete: An equiangular triangle is also _____.

22. If each side of an equiangular triangle measures 4.3 centimeters, find the perimeter of
    this triangle. _____.

Basic Constructions Justified

1. Complete missing statements and reasons in the following proof, which
    justifies the method of construction of an angle congruent to a given angle.

Given: $\angle$ ABC with

$\overline{BD} \cong \overline{BE} \cong \overline{ST} \cong \overline{SR}$

and $\overline{DE} \cong \overline{TR}$

Prove: $\angle$ B $\cong$ $\angle$ S

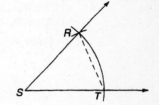

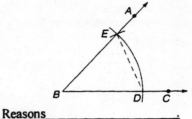

Proof

Statements	Reasons
1. $\angle$ ABC with $\overline{BD} \cong \overline{BE} \cong \overline{ST} \cong \overline{SR}$	1. _____.
2. _____.	2. Given
3. $\triangle$ EBD $\cong$ $\triangle$ RST	3. _____.
4. _____.	4. _____.

2. Construct an isosceles triangle with vertex angle A
    as provided and whose legs are of length ℓ .

A

_____
ℓ

3. Complete missing statements and reasons in the following proof, which
    justifies the method of constructing the bisector of a given angle.

Given: $\angle$ XYZ with $\overline{YM} \cong \overline{YN}$

and $\overline{MW} \cong \overline{NW}$

Prove: $\overline{YW}$ bisects $\angle$ XYZ

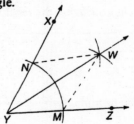

Proof

Statements	Reasons
1. $\angle$ XYZ with $\overline{YM} \cong \overline{YN}$ and $\overline{MW} \cong \overline{NW}$	1. _____.
2. _____.	2. Identity
3. $\triangle$ YMW $\cong$ $\triangle$ YNW	3. _____.
4. _____.	4. CPCTC
5. _____.	5. If ray divides an angle into 2 $\cong$ parts, then it bisects that angle.

Constructing Angles of a Given Measure

4.   a) Construct an angle whose measure is 60^0.
     b) Using the result from part a), construct an
        angle whose measure is 30^0.

5.   a) By construction of perpendicular lines,
        construct an angle whose measure is 90^0.
     b) Using the result from part a), construct an
        angle whose measure is 45^0.

Regular Polygons

6.   a) Use the formula $I = \dfrac{(n-2)180^0}{n}$ to determine the measure
        of each interior angle of a regular hexagon. _____.

     b) Explain how to use the result from 4(a) to construct this angle.

        _____.

7.   a) Use the formula $I = \dfrac{(n-2)180^0}{n}$ to determine the measure
        of each interior angle of a regular octagon. _____.

     b) Explain how to use the result from 5(b) to construct this angle.

        _____.

Inequalities

1. Definition: Where a > b (read "a is greater than b"), there is a _____ number
   p for which a = b + p

2. Because 7 > 5, find the positive number which completes this statement: 7 = 5 + ___.

3. Because 9 = 2 + 7, where 2 and 7 are positive, we know that 9 > ____ and 9 > _____.

Basic Geometric Inequalities (based upon lemmas)

4. According to the Segment-Addition Postulate, AC = AB + BC.
   Then AC > _____ and AC > _____.

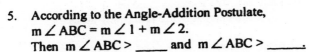

5. According to the Angle-Addition Postulate,
   m ∠ ABC = m ∠ 1 + m ∠ 2.
   Then m ∠ ABC > _____ and m ∠ ABC > _____.

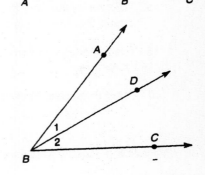

6. The measure of an exterior angle of a triangle is greater than
   the measure of either _____ interior angle.

7. In a right triangle, the largest angle is the _____ angle.

8. Complete (Addition Property of Inequality): If a > b and c > d,
   then _____.

Inequalities in a Triangle

9. Complete: If one side of a triangle is longer than a second side, then the measure of the angle
   opposite the longer side is _____ than the measure of the angle opposite the shorter side.

10. In △ ABC, the 3 part inequality AC > AB > BC is true.
    a)   Which angle of △ ABC is largest?
    b)   Which angle of △ ABC smallest? _____.
    c)   Write a 3 part inequality that compares the measures
         of the angles of this triangle._____.

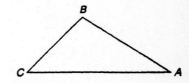

11. Complete: If the measure of one angle of a triangle is greater than the measure of a second angle,
    then the side opposite the larger angle is _____ than the side opposite the smaller angle.

12. Given the measures of angles in △ ABC,
    a)   Find m ∠ C._____.
    b)   Name the longest side. _____.
    c)   Name the shortest side. _____.
    d)   Write a 3 part inequality to compare the lengths
         of the sides of this triangle. _____.

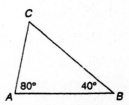

Further Inequality Relationships

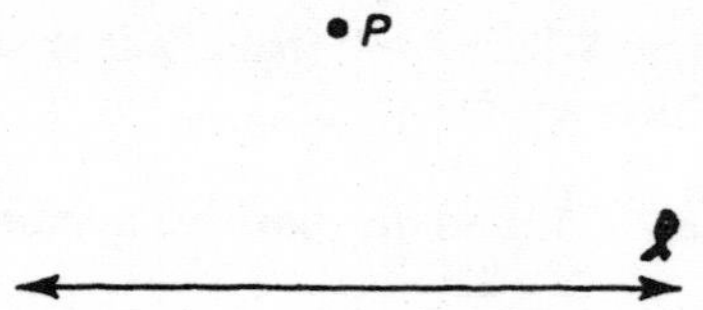

13. What is the shortest distance from point P
    to line ℓ ? _____.

    •P

14. Comparable to the problem 13,
    "The shortest distance from a point not on
    a plane to the plane is the _____.
    segment from point to plane.

    ℓ
    ⟵————————————————⟶

Forms of the Triangle Inequality

15. Triangle Inequality (first form): The sum of lengths of two sides of a triangle is
    greater than the length of the _____ _____.

16. Triangle Inequality(second form): The length of any side of a triangle must lie
    between the _____ and _____ of the lengths of the remaining sides.

17. Using the second form of the Triangle Inequality, can a triangle have these lengths of sides?
    a) 4, 5, and 6 _____   b) 4, 5, and 9 _____   c) 4, 5, and 11 _____.

Proof with Inequality Relationships

18. Complete the missing statements and reasons of the following proof problem.

    Given: AB > CD and BC > DE
    Prove: AC > CE 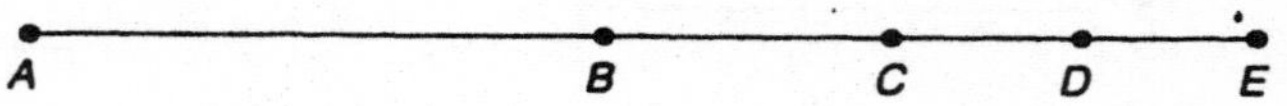
                        Proof

    A———————————B———C——D——E

Statements	Reasons
1. AB > CD and BC > DE	1. _____.
2. AB + BC > CD + DE	2. Addition Property of Inequality
3. AB + BC = AC and CD + DE = CE	3. _____.
4. _____.	4. _____.

The Parallelogram

1. Complete: A parallelogram is a quadrilateral in which both pairs of opposite sides are _____.

2. Complete missing statements and reasons in the proof of the theorem:
   "A diagonal of a parallelogram separates it into two congruent triangles."

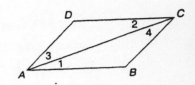

Given: ◻ABCD with diagonal $\overline{AC}$
Prove: △ACD ≅ △CAB
                Proof

Statements	Reasons
1. ◻ABCD with diagonal $\overline{AC}$	1. _____.
2. $\overline{AB}$ ∥ $\overline{CD}$	2. By definition, the opposite sides of a ◻ are parallel.
3. ∠1 ≅ ∠2	3. If 2 lines are cot by a trans, the _____ angles are ≅ .
4. _____.	4. Same as reason 2
5. _____.	5. Same as reason 3
6. $\overline{AC}$ ≅ $\overline{AC}$	6. _____.
7. △ACD ≅ △CAB	7. _____.

Corollaries of the Preceding Theorem

3. Complete: Opposite angles of a parallelogram are _____.

4. Complete: Opposite sides of a parallelogram are _____.

5. Complete: Diagonals of a parallelogram _____ each other.

6. Complete: Consecutive angles of a parallelogram are _____.

Problems Based upon Corollaries

7. In ◻ABCD, AD = 6.3 and AB = 8.7. Find: a) DC _____ b) BC _____.

8. In ◻ABCD, m∠A = 72°. Find: a) m∠C _____ b) m∠B _____.

9. In ◻ABCD, AD = x + 1 and AB = 2x - 5.
   Find: a) x _____ b) AD _____ c) DC _____.

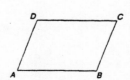

10. In ◻ABCD, m∠A = y + 10 and m∠C = 3y - 90.
    Find: a) y _____ b) m∠A _____ c) m∠B _____.

Exercises 7-10

11. In ▱ ABCD, m ∠ A = y + 10 and m ∠ B = 3y - 90.
    Find: a) y _____ b) m ∠ A _____ c) m ∠ B _____.

Altitude of a Parallelogram

12. In ▱ RSTV, name a line segment that is an altitude to:
    a) side $\overline{VX}$. _____ b) side $\overline{RV}$ _____.

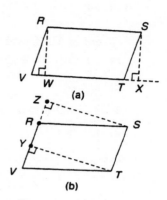

(a)

13. If RW = 6.3 and YT = 8.5, find the height of the
    parallelogram to:
    a) side $\overline{RS}$ _____ b) side $\overline{ST}$. _____.

(b)

Unequal Diagonals of a Parallelogram

14. In ▱ MNPQ, suppose that m ∠ M > m ∠ N.
    Which diagonal ($\overline{QN}$ or $\overline{MP}$) would be longer? _____.

15. In ▱ MNPQ, diagonal $\overline{QN}$ (when drawn) would be longer
    than diagonal $\overline{MP}$. Which angle (∠ P or ∠ Q) would
    be larger? _____.

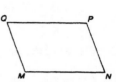

Speed and Direction

16. Give the speed (in mph) and direction (like N 20 ^{0} W) of
    the airplane whose path is indicated. _____.

17. Give the speed (in mph) and direction of the wind
    Whose path is indicated. _____.

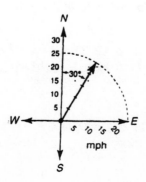

When a Quadrilateral is a Parallelogram

1. Complete this theorem: If two sides of a quadrilateral are both congruent and parallel, then the quadrilateral is a _____.

2. Complete: If both pairs of opposite sides of a quadrilateral are congruent, then the quadrilateral is a _____.

3. Complete: If the diagonals of a quadrilateral _____each other, the quadrilateral is a parallelogram.

4. Complete missing statements and reasons for the proof of the theorem, "If two sides of a quadrilateral are both congruent and parallel, the quadrilateral is a parallelogram.

Given: $\overline{RS} \parallel \overline{VT}$ and $\overline{RS} \cong \overline{VT}$
Prove: RSTV is a parallelogram

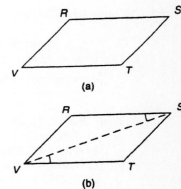

(a)

(b)

Proof

Statements	Reasons
1._____.	1. Given
2. Draw diagonal $\overline{VS}$	2. Through 2 points, there is 1 line.
3. $\overline{VS} \cong \overline{VS}$	3. _____.
4. $\angle RSV \cong \angle SVT$	4. If 2 $\parallel$ lines are cut by a trans, _____.
5. $\triangle RSV \cong \triangle TVS$	5. _____.
6. $\therefore \angle RVS \cong$ _____.	6. CPCTC
7. $\overline{RV} \parallel \overline{ST}$	7. If 2 lines are cut by a trans so that alt. int. $\angle$s are $\cong$, then these lines are parallel.
8. RSTV is a $\square$	8. If both pairs of opposite sides of a quad.

The Kite

5. Complete this definition: A kite is a quadrilateral with two distinct pairs of congruent _____ sides.

6. Complete this theorem: In a kite, one pair of opposite _____ are congruent.

7. In kite ABCD, which pair of angles are congruent? _____

8. If drawn, how would diagonals $\overline{AC}$ and $\overline{BD}$ of kite ABCD be related? _____.

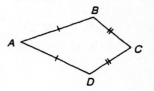

9. Suppose that $\angle B$ is a right angle, AB = 8, and BC = 6.
   Find the length of diagonal $\overline{AC}$. _____.

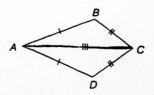

10. Complete missing statements and reasons for the proof of this theorem.
"In a kite, one pair of opposite angles are congruent."

Given: Kite ABCD; $\overline{BC} \cong \overline{CD}$ and $\overline{AD} \cong \overline{AB}$
Prove: $\angle B \cong \angle D$

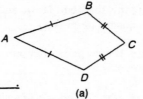
(a)

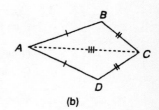
(b)

Proof

Statements	Reasons
1._____ .	1. Given
2. Draw diagonal $\overline{AC}$	2. Through 2 points, there is 1 line.
3. $\overline{AC} \cong \overline{AC}$	3. _____ .
4. $\triangle ACD \cong \triangle ACB$	4. _____ .
5. _____ .	5. _____ .

An Important Theorem

11. Refer to the drawing at the right and complete this theorem:
The line segment that joins the midpoints of two sides of a triangle
is _____ to the third side and has a length equal to
_____ of the length of the third side.

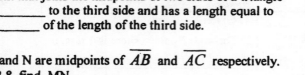

12. In $\triangle ABC$, M and N are midpoints of $\overline{AB}$ and $\overline{AC}$ respectively.
    a) If BC = 12.8, find MN. _____ .
    b) If MN = 5.3, find BC. _____ .
    c) If MN = x + 3, find an expression for BC. _____ .
    d) If MN = x + 3 and BC = 4x − 12, find x. _____ .

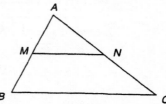

13. In $\triangle RST$, points X, Y, and Z are midpoints of the sides as shown.
    If RS = 18, RT = 24, and ST = 26, find:
    a) XY _____ b) YZ _____ c) XZ _____ .
    d) the perimeter of $\triangle RST$ _____ .

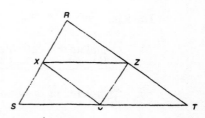

14. In $\triangle RST$, points X, Y, and Z are midpoints of the sides as shown.
    If XY = 7.2, XZ = 6.9, and YZ = 5.1, find:
    a) RS ____ b) RT ____ c) ST ____ .
    d) the perimeter of $\triangle RST$ _____ .

15. When the midpoints of the sides of any quadrilateral are joined in order, the
quadrilateral formed must be a _____ .

The Rectangle

1. Complete: A rectangle is a _____ that has a right angle.

2. In rectangle ABCD, ∠ A is a right angle. Because a rectangle is a type
   of parallelogram, its opposite angles are congruent and m ∠ C = ____.

3. In rect. ABCD, let m ∠ B = m ∠ D = x. Knowing that the sum of the four
   interior angles of any quadrilateral is 360^0 and m ∠ A = m ∠ C = 90^0,
   we know that x + x + 90 + 90 = 360.It follows that,
   "All angles of a rectangle are _____ angles."

4. Complete missing statements and reasons in the proof of the theorem,
   "The diagonals of a rectangle are congruent."

   Given: Rect. MNPQ with diagonals $\overline{MP}$ and $\overline{NQ}$.

   Prove: $\overline{MP} \cong \overline{NQ}$

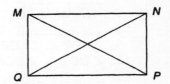

Proof

Statements	Reasons
1. _____	1. _____.
2. MNPQ is a ▱	2. By definition, a rectangle is a ▱ .
3. $\overline{MN} \cong \overline{PQ}$	3. _____
4. $\overline{MQ} \cong \overline{MQ}$	4. _____
5. ∠ s NMQ and PQM are rt. ∠ s	5. _____.
6. _____.	6. All right angles are congruent.
7. △ NMQ ≅ △ PQM	7. _____.
8. _____.	8. _____.

The Square

5. Complete: A square is a rectangle that has two congruent _____ sides.

6. Because a square is a rectangle (and thus a type of parallelogram), it follows that
   "All four sides of a square are _____."

7. Like a rectangle, the diagonals of a square are _____.
   Unlike a rectangle, the diagonals of a square are also _____.

The Rhombus

8. Complete: A rhombus is a _____ with two congruent adjacent sides.

9. Because a rhombus is a type of parallelogram,
    all four sides of rhombus ABCD are _____.

10. Find the perimeter of rhombus ABCD if AB = 5.7 cm. _____.

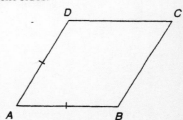

11. Complete the missing statements and reasons in the proof of the following theorem.
"The diagonals of a rhombus are perpendicular."

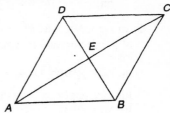

Given: Rhombus ABCD with diagonals $\overline{AC}$ and $\overline{DB}$

Prove: $\overline{AC} \perp \overline{DB}$

Proof

Statements	Reasons
1. _____.	1. _____.
2. ABCD is a ▱	2. By definition, a rhombus is a ▱ .
3. $\overline{DB}$ bisects $\overline{AC}$	3. Diagonals of a ▱ _____.
4. $\overline{AE} \cong \overline{EC}$	4. _____.
5. $\overline{AD} \cong \overline{DC}$	5. All sides of a rhombus are _____.
6. _____.	6. Identity
7. $\triangle ADE \cong \triangle CDE$	7. _____.
8. $\angle DEA \cong \angle DEC$	8. _____.
9. _____.	9. If 2 lines meet to form $\cong$ adj. $\angle$ s, then these lines are $\perp$.

The Pythagorean Theorem

12. The figure at right is a reminder of the Pythagorean Theorem,
"In a right triangle with hypotenuse of length c and legs of
lengths a and b, it follows that _____. "

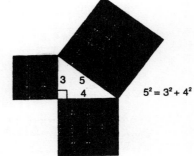

$5^2 = 3^2 + 4^2$

13. What is the length of the diagonal of a rectangle
whose sides have lengths of 3 ft and 4 ft? _____.

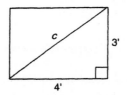

14. What is the length of each side of a rhombus whose
diagonals measure 10 cm and 24 cm? _____.

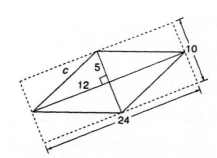

- 42 -

The Trapezoid

1. Definition: A trapezoid is a quadrilateral with exactly two _____ sides.

2. Given that $\overline{HL} \parallel \overline{JK}$ in trapezoid HJKL, name:
   a) the bases _____ b) the legs _____.
   c) two pairs of base angles _____ and _____.

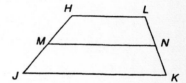

3. If $\overline{HL} \parallel \overline{JK}$ in HJKL, which two sides would have to be
   congruent for HJKL to be an isosceles trapezoid? _____.

4. If the midpoints M and N of the nonparallel sides of trapezoid HJKL are
   joined, then $\overline{MN}$ is called the _____ of HJKL.

5. Intuition suggest that $\overline{MN}$ is _____ to each base of HJKL.

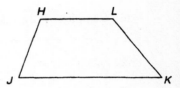

6. A line segment from one vertex of a trapezoid that is perpendicular to the
   opposite base is a(n) _____ of the trapezoid.

Theorems about the Trapezoid

7. Complete: The base angles of an isosceles trapezoid are _____.

8. Complete: The diagonals of a(n) _____ trapezoid are congruent.

9.   Complete: The length of the median of a trapezoid is equal
     to one-half of the _____ of the lengths of the two bases.

Problems involving Trapezoids

10. In trapezoid HJKL with median $\overline{MN}$, suppose that m $\angle$ H = 123^0,
     m $\angle$ K = 72^0, HL = 14, and JK = 22. Find:
       a) m $\angle$ J _____ b) m $\angle$ L _____ c) MN _____.

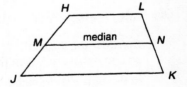

11. In trapezoid HJKL with median $\overline{MN}$, suppose that
     HL = 15 and MN = 23. Find JK. _____.

12. In trapezoid HJKL with median $\overline{MN}$, suppose that
     HL = x, MN = x + 7, and JK = 3x - 6. Find:
       a) x _____ b) HL _____ c) MN _____ d) JK _____.

When a Trapezoid is Isosceles (Theorems)

13. Complete: If a pair of base angles of a trapezoid are _____, then the trapezoid is isosceles.

14. Complete: If the diagonals of a trapezoid are congruent, then the trapezoid is _____.

Proof of a Theorem

15. Complete the missing statements and reasons in the proof of the following theorem: "The base angles of an isosceles triangle are congruent."

Given: Trapezoid RSTV with $\overline{RS} \parallel \overline{VT}$

and $\overline{RV} \cong \overline{ST}$
Prove: ∠V ≅ ∠T

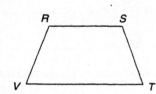 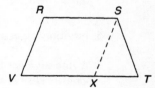

Proof

Statements	Reasons
1. Trap RSTV; $\overline{RS} \parallel \overline{VT}$ and $\overline{RV} \cong \overline{ST}$	1. _____.
2. Draw $\overline{SX} \parallel \overline{RV}$ as shown	2. Parallel Postulate
3. RSVX is a ▱	3. _____.
4. $\overline{RV} \cong \overline{SX}$	4. Opposite sides of a ▱ are _____.
5. $\overline{ST} \cong \overline{SX}$	5. Transitive Prop. of_____.
6. ∠SXT ≅ ∠T	6. If 2 sides of a △ are ≅, then the _____ ∠s are ≅.
7. ∠V ≅ ∠SXT	7. If 2 ∥ lines are cut by a trans., then the _____ ∠s are ≅.
8. _____.	8. _____.

One More Property

16. Consider the theorem, "If 3 (or more) parallel lines intercept congruent segments on one transversal, then they intercept congruent segments on any transversal."

In the figure at the right, a ∥ b ∥ c. Also, $\overline{DE} \cong \overline{EF}$.
Using the theorem, what conclusion can you draw? _____.

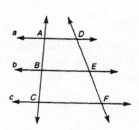

17. In the figure at the right, a ∥ b ∥ c. Also, If DE = 4.7, EF = 4.7, and AB = 4.2, find BC. _____.

Ratios and Rates

1.   Write the ratio in lowest terms. Units are not necessary in the answer!
     a) 6 to 8 _____   b) 12 cm : 20 cm_____   c) 4 inches : 1 foot _____.
                                                    [Note: 1 foot = 12 inches.]

2.   Write these rates in simplest form. Units are necessary!
     a) 200 miles : 10 gallons _____.   b) 56 cents : 4 pencils _____.

Proportions

3. In the proportion $\dfrac{a}{b} = \dfrac{c}{d}$, a and d are the _____ while b and c are the _____.

4.   Solve each proportion for x:

     a) $\dfrac{x}{5} = \dfrac{7}{9}$ _____   b) $\dfrac{x+1}{2} = \dfrac{4}{x-1}$ _____.

     c) $\dfrac{x}{4} = \dfrac{x-2}{3}$ _____   d) $\dfrac{x+3}{5} = \dfrac{4}{x-5}$ _____.

5.   When the second and third terms of a proportion are the same, as in $\dfrac{a}{b} = \dfrac{b}{c}$, the number

     b is called the _____ _____ of a and c.

6.   Find the geometric mean of 4 and 9. _____.

Applications

7.   If an automobile can travel 140 miles on 5 gallons of gasoline, how far can it travel using 7 gallons
     gallons of gasoline? _____.

8.   If two unknown numbers are in the ratio a : b, the numbers can be represented by ax and bx.
     Use the preceding fact to find the measures of two complementary angles that are in the
     ratio 2 : 3. _____ and _____.

9. When 3 unkown numbers have the ratio a : b : c, these numbers can be represented by ax, bx, and cx. Use the preceding fact to find the measures of the three angles of a triangle if the measures of the 3 angles have the ratio 1 : 2: 3. _____ , _____ , and_____ .

Properties of a Proportion

10. One of the claims found in Property 2 of this section is that the means of a true proportion can be interchanged to form another true proportion. By interchanging the means of the proportion $\frac{12}{18} = \frac{2}{3}$, form another true proportion. _____ .

11. One of the claims found in Property 3 takes the form, "If $\frac{a}{b} = \frac{c}{d}$, then $\frac{a+b}{b} = \frac{c+d}{d}$."

    Given the proportion $\frac{6}{8} = \frac{3}{4}$, use this property to write another true proportion. _____ .

Extended Proportions

12. An extended proportion takes the form $\frac{a}{b} = \frac{c}{d} = \frac{e}{f} = \ldots$. Complete this extended proportion

    used in preparing different numbers of servings of some dish.
    $$\frac{2eggs}{3cups} = \frac{4eggs}{6cups} = \frac{?eggs}{9cups} . _____ .$$

13. In $\triangle ABC$ and $\triangle DEF$, it is known that $\frac{AB}{DE} = \frac{AC}{DF} = \frac{BC}{EF}$.

    Using the lengths provided, find x (length of $\overline{DF}$)

    and y (length of $\overline{EF}$).   x = _____   and y = _____ .

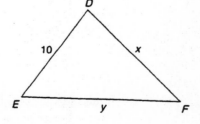

Similar Polygons

1. a) True/False: Any two congruent polygons are also similar polygons. _____.
    b) True/False: Any two similar polygons are also congruent polygons.

2. Given that quad. ABCD and quad. RSTV are similar,
    a) which angle of quad. RSTV corresponds to ∠ A of quad. ABCD? _____.
    b) which side of quad. ABCD corresponds to side $\overline{TV}$ of quad. RSTV? _____.

3. Complete: Two polygons are similar if and only if two conditions are satisfied:
    (i) All pairs of corresponding angles are _____.
    (ii) All pairs of corresponding sides are _____.

4. Which figures must be similar?
    a) any two isosceles triangles? _____   b) any two equilateral triangles? _____.
    c) any two squares? _____   d) any two rectangles? _____.

Problems With Similar Polygons

5. Given that △ ABC ~ △ XTN, suppose that m ∠ A = 92^0 and m ∠ T = 27^0, find:
    a) m ∠ B _____   b) m ∠ X _____   c) m ∠ C _____.

6. Given that △ ABC ~ △ XTN, find x and y that complete
    this extended proportion. $\dfrac{AB}{XT} = \dfrac{x}{TN} = \dfrac{AC}{y}$.
    a) x _____   b) y _____.

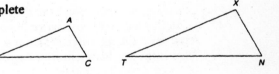

7. Given that △ ABC ~ △ XTN, suppose that AB = 7, AC = 4,   Exercises 5-8
    BC = 8, and XT = 10. Find:
    a) XN _____   b) TN _____.

8. When polygons are similar, their corresponding sides have a constant of proportionality.

    Where △ ABC ~ △ XTN, suppose that AB = 6 and XT = 9. Then AB = $\dfrac{2}{3}$ (XT),

    and it follows that   AC = $\dfrac{2}{3}$ (XN). If XN = 6, find AC. _____.

Further Problems

9. Suppose that quad. ABCD ~ quad. HJKL, with congruent angles as marked.
    If m ∠ A = x, m ∠ C = 2x, and m ∠ K = 3(x − 10), find:
       a) x _____   b) m ∠ K _____.

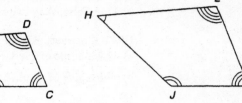

10. Suppose that quad. ABCD ~ quad. HJKL,
    with congruent angles as marked. If AB = 6,
    BC = 4, CD = 5, DA = 7, and HJ = 12, find
    the perimeter of quadrilateral HJKL. _____.

Exercises 9 & 10

Selected Applications

11. On a blueprint, the length of an 18 foot room is represented by a line segment 3.6 inches long.
    By what length (on the blueprint) is the 15 foot width of the room represented? _____.

12. The following problem is based upon shadow reckoning.
    A person 6 feet tall casts a shadow 9 feet in length.
    What is the height of a nearby flagpole that casts
    a 30 foot shadow at that time of day? _____.

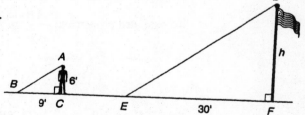

13. It is given that △ ADE ~ △ ABC. If DE = 3, AC = 16,
    and EC = BC, find the length of BC. _____.
    [Hint: There are 2 answers.]

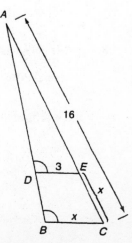

- 48 -

A Method of Proving Triangles Similar

1.   While AAA (Postulate 15) can be used to prove that two triangles are similar, it is better to use the corollary AA because it involves showing that only _____ pairs of angles are congruent.

2.   Are the triangles shown similar if:

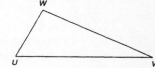
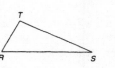

a) $\angle T \cong \angle W$ and $\angle R \cong \angle U$? _____.

b) $m \angle T = m \angle W = 90^0$ and
   $m \angle R = 65^0$ while $m \angle S = 25^0$? _____.

c) $m \angle T = m \angle W = 90^0$ and
   $m \angle R = 66^0$ while $m \angle S = 26^0$? _____.

Proof Using AA

3.   Complete the missing statements and reasons in the following proof

Given: $\overline{AB} \parallel \overline{DE}$
Prove: $\triangle ABC \sim \triangle EDC$

Proof

Statements	Reasons
1. _____	1. _____
2. $\angle A \cong \angle E$	2. If 2 $\parallel$ lines are cut by a trans, _____
3. $\angle 1 \cong \angle 2$	3. _____
4. _____	4. AA

4.   Once triangles have been proven congruent, we can show that:
a) corresponding _____ are proportional by the reason CSSTP; and
b) corresponding _____ are congruent by the reason CASTC.

5.   Complete the missing statements and reasons in the following proof

Given: $\angle ADE \cong \angle B$

Prove: $\dfrac{DE}{BC} = \dfrac{AE}{AC}$

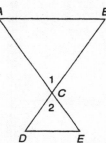

Proof

Statements	Reasons
1. _____	1. _____
2. $\angle A \cong \angle A$	2. _____
3. $\triangle ADE \sim \triangle ABC$	3. _____
4. _____	4. CSSTP

6.   In problem 5, suppose that you were asked to prove that $DE \cdot AC = BC \cdot AE$. That would add one statement (step 5) to the proof above. The reason that would justify the fact is called the _____-_____ Product Property.

7.   Complete this theorem: The lengths of the corresponding altitudes of two similar triangles have the same ratio as the lengths of two corresponding _____.

Further Methods for Showing that Triangles are Similar

8.   In addition to the AA method for proving triangles similar are:
(i) the method _____ used when included angles of the triangles are formed by pairs of sides whose lengths are proportional.
(ii) the method _____ used when the lengths of all three pairs of corresponding sides are proportional.

9.   As a point of clarification, SAS and SSS are methods used to prove that two triangles are _____ while SAS ~ and SSS ~ are used to prove that triangles are _____.

10.   Which method (AA, SAS ~ , or SSS ~) proves that $\triangle$ HJK is similar to $\triangle$ FGK if:
a) $\angle$ J and $\angle$ G are right angles. _____.
b) KG = 2(KJ) and FK = 2(HK) _____.
c) $\dfrac{KG}{KJ} = \dfrac{3}{2}, \dfrac{FK}{HK} = \dfrac{3}{2}$, and $\dfrac{FG}{HJ} = \dfrac{3}{2}$. _____.

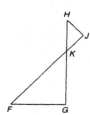

Further Applications

11.   In the figure, $\dfrac{DG}{DE} = \dfrac{DH}{DF}$. If m $\angle$ E = x, m $\angle$ DGH = 2x – 80, and m $\angle$ F = x – 45, find:
a) m $\angle$ E _____   b) m $\angle$ F _____   c) m $\angle$ D _____.

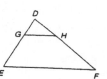

12.   Complete the missing statements and reasons in the following proof

Given: $\dfrac{DE}{BC} = \dfrac{AE}{AC} = \dfrac{AD}{AB}$
Prove: $\angle$ ADE $\cong$ $\angle$ B

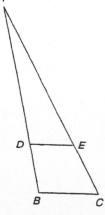

Proof

Statements	Reasons
1. _____	1. _____
2. $\triangle$ ADE ~ $\triangle$ ABC	2. _____
3. _____	3. CASTP

An Important Theorem

1. Complete: The altitude to the hypotenuse of a right triangle separates the right triangle into two right triangles that are _____ to each other and to the original right triangle.

2. Consider the theorem above and the figures below. To prove that $\triangle ADC$ is similar to $\triangle ACB$, we use the method AA. By the reason _____, we know that $\angle A \cong \angle A$. Also, $\angle D$ of $\triangle ADC$ and $\angle C$ of $\triangle ACB$ are both _____ angles; thus, these angles are congruent.

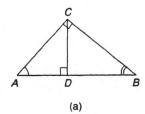
(a)

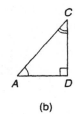
(b)

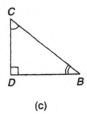
(c)

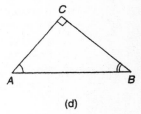
(d)

Proof of the Pythagorean Theorem

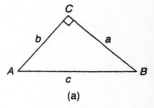
(a)

3. Consider the figures at the right. From the fact that $\triangle ABC \sim \triangle CBD$, we know that $\dfrac{c}{a} = \dfrac{a}{y}$ so that $a^2 =$ _____. Because $\triangle ABC \sim \triangle ACD$, we also know that $\dfrac{c}{b} = \dfrac{b}{x}$ so that $b^2 =$ _____. By substitution, it follows that $a^2 + b^2 = cy + cx = c(y + x)$. By the Segment-Addition Postulate and figure (b), $y + x =$ _____. By substitution, the statement $a^2 + b^2 = c \cdot c$ or $a^2 + b^2 = c^2$.

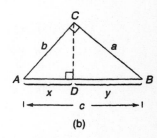
(b)

4. In $\triangle RST$ with right $\angle S$, find:
    a) c if a = 4 and b = 6 _____   b) b if a = 12 and c = 13 _____.

The Converse of the Pythagorean Theorem

5. Complete: Let a, b, and c be the lengths of the 3 sides of a triangle, where c is the length of the longest side. If $c^2 = a^2 + b^2$, then the triangle is a _____ triangle with the right angle opposite the side of length _____.

6. Given the following lengths of sides of a triangle, is the triangle a right triangle?
    a) a = 5, b = 12, c = 13? _____   b) a = 7, b = 9, c = 10? _____.   c) a = $\sqrt{2}$, b = $\sqrt{3}$, c = $\sqrt{5}$? _____.

Application of the Pythagorean Theorem

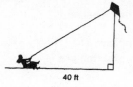

7. A strong wind holds a kite 30 feet above the earth in a
   position 40 feet across the ground. How much string
   does the girl have out to the kite? _____.

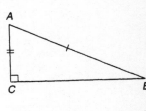

8. A second application of the Pythagorean Theorem is that it enables us to prove the
   HL method for proving that two right triangles are congruent. In the
   drawing at the right, we know that AB = DE = c and AC = EF = a.
   Using the Pythagorean Theorem, we find the lengths CB and DF in
   terms of a and c; in particular, CB = DF = _____. By SSS,
   the triangles are congruent.

Pythagorean Triples

9. Complete this definition: A Pythagorean Triple is a set of three natural numbers (a,b,c)
   for which _____.

10. Which of the triples shown below are Pythagorean triples?
    a) (3,4,5)? _____ b) (4,5,6)? _____ c) (0.3,0.4,0.5)? _____ d) (20,21,29)? _____.

11. Let p and q be natural numbers with p > q. Then (a,b,c) is a Pythagorean Triple enerated
    by the formulas: $a = p^2 - q^2$, b = 2pq, and $c = p^2 + q^2$. Find the Pythagorean Triple for which:
    a) p = 2 and q = 1. _____   b) p = 3 and q = 2. _____.

An Extension of the Converse

12. Complete: Let a, b, and c represent the lengths of the three sides of a triangle, with
    c being the length of the longest side.
    i) If $c^2 = a^2 + b^2$, the triangle is a(n) _____ triangle.
    ii) If $c^2 < a^2 + b^2$, the triangle is a(n) _____ triangle.
    iii) If $c^2 > a^2 + b^2$, the triangle is a(n) _____ triangle.

13. What type of triangle (acute, right, or obtuse) has these lengths for its three sides?
    a) a = 3, b = 4, c = 6? _____ b) a = 3, b = 4, c = 5? _____ c) a = 4, b = 5, c = 6? _____.

The 45^0-45^0-90^0 Triangle

1. In any triangle with 2 congruent angles, the sides opposite these angles are _____.
   Thus, the sides opposite the 45^0 angles of a 45^0-45^0-90^0
   triangle are _____. If the length of the side opposite
   one 45^0 angle measures a, then the length of the side opposite
   the second 45^0 angle has measure _____.

2. If both legs of a 45^0-45^0-90^0 triangle have length a, then the
   length c of the hypotenuse can be found by solving the
   equation $c^2 = a^2 + a^2$ for c. Solve for c.   _c =_ _____.

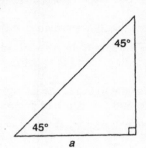

3. Complete the 45^0-45^0-90^0 Theorem: In a triangle whose angles measure 45^0,45^0, and 90^0, the
   hypotenuse has a length equal to the product of _____ and the length of either of the congruent legs.

For Exercises 4 and 5, use the theorem above.

4. Given: $\triangle ABC$ is a 45^0-45^0-90^0 triangle as shown.
   a) If AC = 6, find BC _____ and AB _____.

   b) If AB = $8\sqrt{2}$, find AC _____ and BC _____.

5. Given: $\triangle DEF$ is a 45^0-45^0-90^0 triangle as shown.
   a) If EF = $3\sqrt{2}$, find DE ____ and DF _____.

   b) If DF = 4, find DE ____ and EF _____.

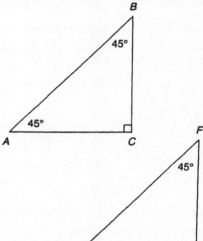

6. The figure shown at the right is a geometric representation
   of the _____ theorem.

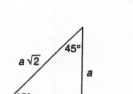

7. Repeat problem 4a, but find the length of the hypotenuse $\overline{AB}$ by using the Pythagorean Theorem
   and the fact that AC = 6 and BC = 6. _____.

The 30^0-60^0-90^0 Triangle

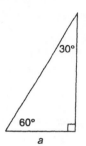
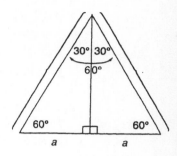

8. Use the figure provided at the right. If the length
    of the shorter leg of a 30^0-60^0-90^0 triangle has
    length a, then the hypotenuse must have length _____.

9. Use the figure at the far right and the Pythagorean Theorem
    To find the length of the longer leg in terms of a. That is,
    solve the equation $(2a)^2 = a^2 + b^2$ for b. __b =__ _____.

10. Complete the 30^0-60^0-90^0 Theorem. In a triangle whose angles measure 30^0, 60^0, and 90^0,
    The hypotenuse has a length equal to twice the length of the _____ leg and the length
    of the longer leg is the product of _____ and the length of the shorter leg.

For Exercises 11-13, use the theorem above.

11. Given: $\triangle$RST is a 30^0-60^0-90^0 triangle as shown.
    a) If RS = 8, find RT _____ and ST _____.

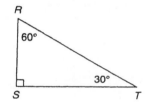

    b) If RT = 12, find RS _____ and ST ____.

12. Given: $\triangle$XYZ is a 30^0-60^0-90^0 triangle as shown.
    a) If YZ = $10\sqrt{3}$, find XY ____ and XZ _____.

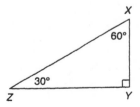

    b) If XY = $2\sqrt{3}$, find YZ ____ and XZ ____.

13. The figure shown at the right is a geometric representation
    of the _____ theorem.

Segments Divided Proportionally

1. In the figure, $\overline{AC}$ and $\overline{DF}$ are divided proportionally

    if $\dfrac{AB}{DE} = \dfrac{BC}{EF}$ or if $\dfrac{AB}{BC}$ equals _____ .

2. In the figure, $\overline{AC}$ and $\overline{DF}$ are divided proportionally.
    If AB = 5, BC = 3, and DE = 7, find EF. _____ .

3. Complete: If a line is parallel to one side of a triangle and intersects the other two sides,
    it divides these sides _____ .

4. Sketch of the proof for the theorem in problem 3:

    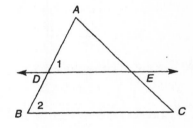

    Given: $\triangle$ ABC with $\overline{DE} \parallel \overline{BC}$ as shown

    Prove: $\dfrac{AB}{AD} = \dfrac{AC}{AE}$

    Proof: Because $\overline{DE} \parallel \overline{BC}$, $\angle 1 \cong \angle 2$.
    With $\angle A \cong \angle A$ by the reason _____ ,
    $\triangle$ ADE ~ $\triangle$ ABC by _____ .

    Then $\dfrac{AB}{AD} = \dfrac{AC}{AE}$ by CSSTP. In turn, $\dfrac{AB - AD}{AD} = \dfrac{AC - AE}{AE}$.

    But AB − AD = _____ and AC − AE = _____, so the preceding
    proportion can be written _____ .

    Exercise 3-6

5. Given that $\overline{DE} \parallel \overline{BC}$, AD = 6, DE = 4, and EC = 5, find AE. _____ .

6. Given that $\overline{DE} \parallel \overline{BC}$, AD = x + 2, DB = x, AE = x + 4, and EC = x + 1,
    find x. _____ .

7. Complete: When three (or more) parallel lines are cut by a transversal, the transversals are
    divided proportionally by the _____ lines.

8. In the figure $\ell_1 \parallel \ell_2\ \ell_3$, AC = 12, BC = 7, and DE = 6.
    Find EF. _____ .

    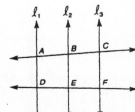

The Angle-Bisector Theorem

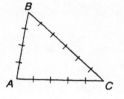
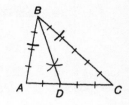

9. In △ABC, AB = 4, BC = 6, and AC = 5. Given that $\overline{BD}$ bisects ∠ABC, use measurements from the drawing to verify the proportion $\dfrac{AB}{BC} = \dfrac{AD}{DC}$ _____.

10. Complete: If a ray bisects one angle of a triangle, then it divides the opposite side into segments whose lengths are proportional to the lengths of the two sides that form the _____ angle.

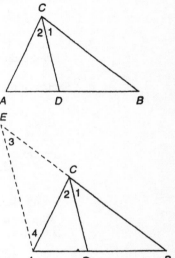

11. Complete this sketch of the proof of the theorem above. This proof requires the use of auxiliary lines.

   Given: △ABC in which $\overline{CD}$ bisects ∠ACB.

   Prove: $\dfrac{AD}{AC} = \dfrac{DB}{CB}$

   Proof: First we draw $\overline{EA}$ _____ to $\overline{CD}$.

   Side _____ of △ABC is extended to meet $\overline{EA}$.

   Because $\overline{CD} \parallel \overline{EA}$ in △_____, it follows that $\dfrac{EC}{AD} = \dfrac{CB}{DB}$ (*).

   Because $\overline{CD} \parallel \overline{EA}$, it follows that ∠2 ≅ ∠_____ (alt. int ∠s).

   Because $\overline{CD} \parallel \overline{EA}$, it follows that ∠1 ≅ ∠_____(corr. ∠s)

   But ∠1 ≅ ∠2 is Given, so ∠3 ≅ ∠4 by the Transitive Property of Congruence. Then $\overline{AC} \cong \overline{EC}$ because these sides lie opposite congruent angles in △_____. With AC = EC, the starred (*) proportion becomes.
   _____ or _____ by inversion

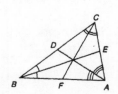

12. In △XYZ, $\overline{YW}$ bisects ∠XYZ. If XY = 4, YZ = 6, and XW = 3, find WZ. _____.

13. In △XYZ, $\overline{YW}$ bisects ∠XYZ. If XY = 6, YZ = 8, and XZ = 7, find WZ. _____.

Ceva's Theorem

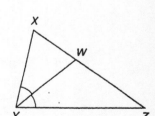

14. According to Ceva's Theorem. $\dfrac{BF}{FA} \cdot \dfrac{AE}{EC} \cdot \dfrac{CD}{DB} = 1$.

   If BF = 5, FA = 4, AE = 3, and EC = 4, find the ratio $\dfrac{CD}{DB}$. _____.

Basic Terminology for Circles

1. In ⊙Q, $\overline{QS}$ is a _____ while $\overline{QW}$, $\overline{QV}$, and $\overline{QT}$ are _____.
   $\overline{SW}$ is a _____ of ⊙Q while $\overline{WT}$ is a
   _____ because it passes through the center Q.

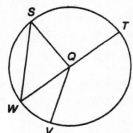

2. In ⊙Q, suppose that $\overline{SQ} \perp \overline{WT}$. If QV = 5, find:
   a) QS ____   b) WT ____   c) SW ____ .

3. Congruent circles have the same length of _____ .
   Concentric circles have the same _____ .

Central Angles

4. In ⊙O, $\overline{MP}$ and $\overline{QN}$ are diameters. Because the vertex of ∠1 is center O
   and the sides are radii, ∠1 is a _____ angle. For the arcs shown, $\overparen{NP}$
   is a _____ arc while $\overparen{NPQ}$ is a _____ and
   $\overparen{NPQM}$ is a _____ arc.

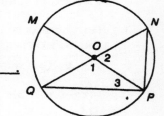

5. Complete (Central angle Postulate): In a circle, the degree measure
   of a central angle is equal to the degree measure of its intercepted _____ .

6. In ⊙O, $\overline{MP}$ and $\overline{QN}$ are diameters. If m $\overparen{NP}$ = 58°, find:
   a) m∠2 ____   b) m∠1 ____ .   c) m $\overparen{QP}$ ____   d) m∠3 ____ .

Exercises 4 & 6

Further Properties

7. Fill in the blanks for the paragraph proof of the theorem,
   "A radius that is perpendicular to a chord bisects the chord."

   Given: In ⊙O, radius $\overline{OD} \perp \overline{AB}$ at point C
   Prove: $\overline{OD}$ bisects $\overline{AB}$.

   Proof: In ⊙O, it is Given that _____ .
   We draw auxiliary radii $\overline{OA}$ and ____ as shown. Now
   $\overline{OA} \cong \overline{OB}$ because "All radii of a circle are _____ ."
   With right ∠1 and right ∠2 and _____ by Identity,
   it follows that △OCA ≅ △OCB by ____ . Then $\overline{AC} \cong \overline{CB}$
   by CPCTC. By definition, $\overline{OD}$ bisects $\overline{AB}$.

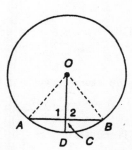

8. In ⊙O, radius $\overline{OD} \perp \overline{AB}$ at point C. If AB = 8
   and OB = 5, find OC. _____.

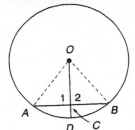

9. Use the Arc-Addition Postulate to complete:

   m $\overset{\frown}{AD}$ + m $\overset{\frown}{DB}$ = _____.

10. For two arcs to be congruent, the arcs must have the same measure and these arcs
    must be parts of a circle or of _____ circles.

Inscribed Angles

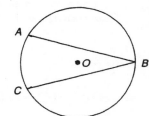

11. In ⊙O, $\overline{AB}$ and $\overline{CB}$ are _____. With the vertex of
    of ∠B on ⊙O, ∠B is a(n) _____ angle. For ∠B,
    the intercepted arc is _____ .

12. Complete: The measure of an inscribed angle of a circle is _____.
    the measure of its intercepted arc.

13. Fill in the blanks for the paragraph proof of Case 1 of the theorem in problem 12.
    "A radius that is perpendicular to a chord bisects the chord."

    Given: ⊙O with inscribed ∠RST and diameter $\overline{TS}$

    Prove: m∠S = $\dfrac{1}{2}$ m $\overset{\frown}{RT}$.

    Proof: ⊙O with inscribed ∠RST and diameter $\overline{TS}$. Draw radius

    _____ . Because ∠ROT is a central angle, m∠ROT = m $\overset{\frown}{RT}$.
    With $\overline{OR} \cong \overline{OS}$, it follows that m∠R = _____ in △ROS.
    Because ∠ROT is an exterior angle of △ROS, m∠ROT = ____ + ___.
    By substitution, m∠ROT = m∠S + m∠S = 2(m∠S).
    In turn, 2(m∠S) = m $\overset{\frown}{RT}$ and m∠S = _____.

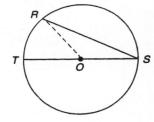

14. In the figure shown, m $\overset{\frown}{AC}$ = 54^0 and m $\overset{\frown}{ED}$ = 38^0. Find:
    a) m∠B _____   b) m∠O _____.

15. In the figure shown, m∠B = 29^0 and m∠O = 36.7^0. Find:
    a) m $\overset{\frown}{AC}$ _____   b) m $\overset{\frown}{ED}$ _____.

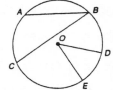

Further Properties

16. Complete (See figure at right): An angle inscribed in a semicircle
    is a _____ angle.

17. Complete: If two inscribed angles intercept the same arc,
    these angles are _____.

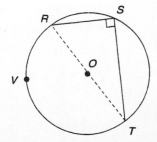

- 58 -

Tangents and Secants

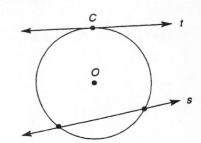

1. Because line t touches ⊙ O at one point, line t is
a(n) _____ to ⊙ O. Point C is called the point
of contact or point of _____.

2. Because line s intersects ⊙ O at two points, line s is
a(n) _____ to ⊙ O.

Inscribed and Circumscribed Polygons (and Circles)

3. A polygon is _____ in a circle if its vertices lie
on the circle and its sides are _____ of the circle.
In the figure, the inscribed quadrilateral RSTV is also
known as a cyclic _____.

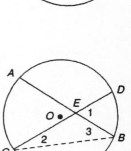

4. Because RSTV is inscribed in ⊙ Q, the circle is said to be
_____ about RSTV.

5. Complete: If a quadrilateral is cyclic, then its opposite angles are _____.

6. Complete: A polygon is circumscribed about a circle if its sides are _____ to the circle.

More Angles in the Circle

7. Complete: The measure of an angle formed by two chords that intersect within a circle is one-half
the sum of measures of the arcs intercepted by the angle and its _____ angle.

8. Fill the blanks to complete the proof of the theorem above in Exercise 7.

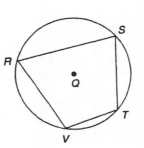

Given: In ⊙ O, chords $\overline{AB}$ and $\overline{CD}$ intersect at E.

Prove: $m \angle 1 = \frac{1}{2} (m \overarc{AC} + m \overarc{DB})$

Proof: In ⊙ O, chords $\overline{AB}$ and $\overline{CD}$ intersect at E.

Draw chord $\overline{CB}$. Because ∠1 is an exterior angle of

△CBE, m∠1 = _____ + _____. But m∠2 = $\frac{1}{2}$ m $\overarc{DB}$

and m∠3 = _____. Substitution into the starred statement

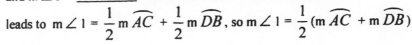

leads to m∠1 = $\frac{1}{2}$ m $\overarc{AC}$ + $\frac{1}{2}$ m $\overarc{DB}$, so m∠1 = $\frac{1}{2}$ (m $\overarc{AC}$ + m $\overarc{DB}$)

9. In the figure for Exercise 8, m $\overarc{AC}$ = 68° and m $\overarc{DB}$ = 44°.
Find: a) m∠1 _____.   b) m∠CEB _____.

10. Complete: The radius drawn to a tangent at its point of tangency is _____ to
the tangent.

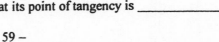

Angles Formed by Tangent and Chord

10. Complete: The measure of an angle formed by a tangent and a chord drawn to the point of tangency
is _____ the measure of the intercepted arc.

11. In the drawing, $\overline{AC}$ is tangent to $\odot O$ and $\overline{DB}$ is a diameter.
If m $\overset{\frown}{EB}$ = 82^0, find: a) m $\angle$ 2 _____   b) m $\angle$ 1 _____.
c) m $\angle$ ABD ____   d) m $\overset{\frown}{DE}$ _____.

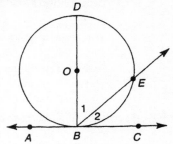

Angles With Vertex in Exterior of Circle

12. Complete: The measure of an angle formed when two secants intersect at a point outside the
circle is one-half the _____ of the measures of the two intercepted arcs.

13. Fill the blanks to complete the proof of the theorem found in Exercise 12.

Given: In the circle, secants $\overline{AC}$ and $\overline{DC}$ intersect at C.

Prove: m $\angle$ C = $\frac{1}{2}$ (m $\overset{\frown}{AD}$ - m $\overset{\frown}{BE}$)

Proof: In the circle, secants $\overline{AC}$ and $\overline{DC}$ intersect at C.
Draw $\overline{BD}$ to form $\triangle$ BCD. Because $\angle$ 1 is an exterior angle of
$\triangle$ BCD, m $\angle$ 1 = _____ + _____. Then m $\angle$ C = m $\angle$ 1 - _____.(*)
But m $\angle$ 1 = $\frac{1}{2}$ m $\overset{\frown}{AD}$ and m $\angle$ D = _____. $\overset{\frown}{DB}$
Substitution into the starred statement leads to
m $\angle$ 1 = $\frac{1}{2}$ m $\overset{\frown}{AD}$ - $\frac{1}{2}$ m $\overset{\frown}{BE}$, so $\frac{1}{2}$ (m $\overset{\frown}{AD}$ - m $\overset{\frown}{BE}$)

14. In the figure for Exercise 13, find:
a) m $\angle$ C; m $\overset{\frown}{AD}$ = 117^0 and m $\overset{\frown}{BE}$ = 39^0 ____b) m $\overset{\frown}{AD}$;m $\angle$ C = 35^0 and m $\overset{\frown}{BE}$ = 37^0.____

15. Two other types of angles whose measures are one-half the difference between intercepted arcs
are formed by: i) a _____ and a _____ or ii) two _____.

16. In the drawing at right, find:
a) m $\angle$ L if m $\overset{\frown}{JH}$ = 125^0 and m $\overset{\frown}{JK}$ = 47^0 _____.

b) m $\overset{\frown}{JK}$, if m $\angle$ L = 38^0 and m $\overset{\frown}{JH}$ = 3(m $\overset{\frown}{JK}$) _____.

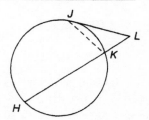

17. In the drawing at right, find m $\angle$ 1 if m $\overset{\frown}{ACB}$ = 235^0. _____.

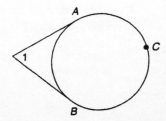

Three Related Theorems

1. Complete: If a line through the center of a circle is perpendicular to a chord, then it bisects the _____ and its _____.

2. Complete: If a line through the center of a circle bisects a chord (other than a diameter), then it is _____ to the chord.

3. Complete: The perpendicular _____ of a chord of a circle contains the center of the circle.

4. In the figure, ⊙O has a radius of length 10.
    Also, $\overline{OE} \perp \overline{CD}$ at B and OB = 6. Find:
    a) CB _____   b) CD _____.

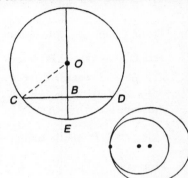

Tangent Circles and Tangents to Circles

5. The circles shown at the right are _____ tangent.

6. How many common tangents do the circles have?
    a) _____   b) _____   c) _____.

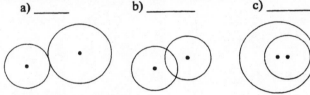

Tangent Segments to Circles

7. Complete: The tangent segments to a circle from an external point are _____.

8. Fill in the blanks to complete the proof of the theorem in Exercise 7.

    Given: $\overline{AB}$ and $\overline{AC}$ are tangents to ⊙O from point A
    Prove: $\overline{AB} \cong \overline{AC}$
    Proof: Draw $\overline{BC}$, which along with tangents $\overline{AB}$ and $\overline{AC}$

    form Δ _____. Now m ∠ B = $\dfrac{1}{2}$ m $\overparen{BC}$ and m ∠ C = _____

    Then ∠B ≅ ∠C. In ΔABC, _____ since these angles opposite congruent angles of a triangle.

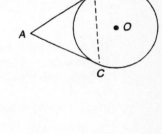

9. In the figure, suppose that AP = 5.2, CN = 4.3, and MB = 7.1.
    Find: a) AM _____ b) AC _____ c) BC _____ d) AB _____.

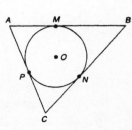

10. In the figure for Exercise 9, suppose that AB = 14, BC = 16, and AC =12.
    Let AM = x, CP = y, and BN = z. Write 3 equations in variables x, y, and z.
    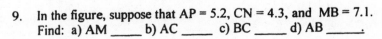
    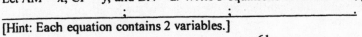
    [Hint: Each equation contains 2 variables.]

Lengths of Segments in a Circle

11. Complete: If two chords intersect within a circle, the product of the lengths of segments (parts) of one chord is equal to the _____ of the lengths of segments of the other chord.

12. Fill the blanks to complete the proof of the theorem in Exercise 11.

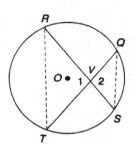

Given: $\odot$ O and intersecting chords $\overline{RS}$ and $\overline{TQ}$
Prove: $RV \cdot VS = TV \cdot VQ$

Proof: Given $\odot$ O and chords $\overline{RS}$ and $\overline{TQ}$, we draw $\overline{RT}$ and $\overline{QS}$ to form $\triangle RTV$ and $\triangle$ _____. In the triangles, $\angle 1 \cong \angle 2$ because these are _____ angles. Also, $\angle R \cong \angle$ ____ because these inscribed angles intercept the same arc. Then $\triangle RTV \sim \triangle QSV$ by the reason _____.

Using CSSTP, it follows that $\dfrac{RV}{VQ} = \dfrac{TV}{VS}$. By the Means-Extremes

Property, _____.

13. Use the figure for Exercise 12.
   a)   Find VS if RV = 8, TV = 7, and VQ = 6. _____.

   b)   Find RV if TV = 6, VS = 4, and RS = 11. _____.

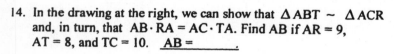

14. In the drawing at the right, we can show that $\triangle ABT \sim \triangle ACR$ and, in turn, that $AB \cdot RA = AC \cdot TA$. Find AB if AR = 9, AT = 8, and TC = 10.   **AB =** _____.

15. Based upon the drawing in Exercise 14, complete this theorem:
   If two secant segments are drawn to a circle form an external point, then the products of the length of each secant with the length of its _____segment are equal.

16. In the drawing at the right, it can be shown that $\triangle TVW \sim \triangle TXW$
   and in turn that $(TV)^2 = TW \cdot TX$.
   a) Find TV if TW = 18 and TX = 8. _____.

   b) Find TX if $TV = 2\sqrt{15}$, TX = x, and TW = x + 4 . _____.

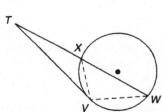

17. Based upon the drawing in Exercise 16, complete this theorem:
   If a tangent segment and a secant segment are drawn to a circle form an external point, then the _____ of the length of tangent equals the product of the length of the secant with the length of its external segment..

Construction of Tangents to Circles

1. Complete: The line that is perpendicular to the radius of a circle at its endpoint on the circle is a _____ to the circle.

2. Use the theorem in Exercise 1 to construct the tangent to ⊙ P at point X.

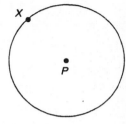

3. The construction of the tangent(s) to ⊙ Q from external point E can be justified by the fact that ∠ ETQ is a right angle. Then $\overline{ET} \perp \overline{TQ}$.

   According to the theorem in Exercise 1, $\overline{ET}$ is a(n) _____ to ⊙ Q.

Fig 6.56 c

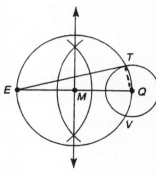

Inequalities in a Circle

4. Complete: In a circle (or in congruent circles) containing two unequal central angles, the larger central angle corresponds to the larger _____ arc.

5. Fill the blanks in the proof of the theorem in Exercise 4.

   Given: ⊙ O with central ∠ s 1 and 2; m ∠ 1 > m ∠ 2

   Prove: m $\overarc{AB}$ > m $\overarc{CD}$

   Proof: In ⊙ O, m ∠ 1 > _____. But m ∠ 1 = m $\overarc{AB}$ and m ∠ 2 = _____. By substitution, _____.

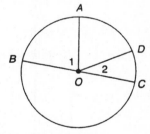

6. The converse of the theorem in Exercise 4 is also true. Using the figure in Exercise 5 and the following information, draw a conclusion.

   a) m $\overarc{AB}$ > m $\overarc{CD}$ _____. b) m ∠ 1 = 72^0 and m ∠ 2 = 36^0 _____.

7. Complete: In a circle (or in congruent circles) containing two unequal chords, the _____ chord is at a greater distance from the center of the circle.

8. The converse of the theorem in Exercise 7 is also true. Complete the converse. In a circle (or in congruent circles) containing two unequal chords, the chord that is nearer the center of the circle has the _____ length.

9. In the drawing at the right, RP = 8, PS = 7, and TP = 4. Use the theorem of Exercise 8 to determine which chord shown is:
   a) longest? _____ b) shortest? _____.

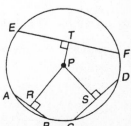

- 63 -

10. Fill in the blanks for the proof of the theorem of Exercise 8.

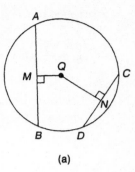

Given: $\odot$ Q with chords $\overline{AB}$ and $\overline{CD}$;

$\overline{QM} \perp \overline{AB}$, $\overline{QN} \perp \overline{CD}$ and QM < QN

Prove: AB > CD

Proof: In the given figure, radii of length r are drawn as shown .

Now $\overline{QM}$ is the perpendicular-bisector of _____ while _____ is

the perpendicular-bisector of $\overline{CD}$. Let MB = b and NC = d.
With right angles as indicated, we apply the Pythagorean Theorem
to show that $r^2 = a^2 + b^2$ in $\triangle$ QMB while $r^2 = $ _____ in
$\triangle$ QNC. Then $b^2 = r^2 - a^2$ and $d^2 = $ _____ . With QM < QN,
we know that a < c and, in turn, that $a^2 < c^2$. Multiplying by –1,
the last inequality reverses order to become _____. By addition,
$r^2 - a^2 > r^2 - c^2$; by substitution, $b^2 > d^2$ which implies that
_____ . Then 2b > 2d. But AB = 2b and CD = 2d. Then AB > CD.

(a)

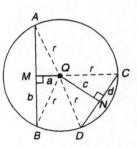

11. In the figure for Exercise 10 is $\odot$ Q Suppose that the length of the radius is 5.
If QM = 3 and QN = 4, find:
a) AM _____ b) AB _____ c) NC _____ d) _____ .
[Hint: Use the Pythagorean Theorem.]

12. In the drawing for Exercise 10, suppose that QN = 5 and QM = 4. State an inequality
that compares the lengths of chords $\overline{AB}$ and $\overline{CD}$. _____ .

13. Complete: In a circle (or congruent circles) containing two unequal chords, the
_____ chord corresponds to the greater minor arc.

14. In the figure, AB = 6.8 and CD = 4.9. State an inequality that
compares m $\overset{\frown}{AB}$ and m $\overset{\frown}{CD}$. _____ .

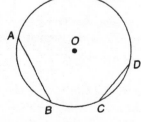

15. Complete: In a circle (or congruent circles) containing two unequal minor arcs, the greater
minor arc corresponds to the _____ chord related to these arcs.

16. In the figure, m $\overset{\frown}{MN}$ = 58^0 and m $\overset{\frown}{PQ}$ = 29^0. Write an inequality that
compares the lengths of chords $\overline{MN}$ and $\overline{PQ}$. _____ .

Locus of Points

1. Definition: A _____ is the set of all points (and only those points) that satisfy a given
    condition or set of conditions.

2. Draw and describe "the locus of points in a plane that
    are at a distance of 1 inch from given point P."   · P
    The locus is the _____.

    _____.

3. In Exercise 2, suppose that the locus were to be determined
    in space. What type of geometric figure is characterized? _____.

4. Draw and describe "the locus of points in a plane that
    are at a distance of 1 centimeter from a **given line** ℓ ."
    The locus is _____.

    _____.

An Important Theorem

5. Complete: The locus of points in a plane that are equidistant from the
    sides of an angle is _____.

6. Fill in the blanks for the 2-part proof of the theorem in Exercise 5.
    i)   If a point is on the angle-bisector, then
            it is equidistant from the sides of the angle.

    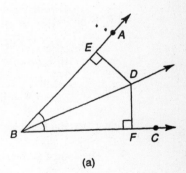

    Given: $\overline{BD}$ bisects $\angle$ ABC;

    $\overline{DE} \perp \overline{BA}$ and $\overline{DF} \perp \overline{BC}$

    Prove: $\overline{DE} \cong \overline{DF}$

    Proof: $\overline{BD}$ bisects $\angle$ ABC, so $\angle$ ABD $\cong$ $\angle$ _____.
    $\overline{DE} \perp \overline{BA}$ and $\overline{DF} \perp \overline{BC}$, so $\angle$ DEB and $\angle$ DFB
    are _____ angles. Then $\angle$ DEB $\cong$ _____. By
    identity, _____. Then $\triangle$ DEB $\cong$ _____ by AAS.
    Then _____ by CPCTC.

    (a)

    ii)   If a point is equidistant from the sides of an angle,
            then it lies on the angle-bisector.

    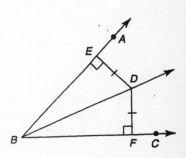

    Given: $\angle$ ABC so that $\overline{DE} \perp \overline{BA}$ and
    $\overline{DF} \perp \overline{BC}$; $\overline{DE} \cong \overline{DF}$

    Prove: $\overline{BD}$ bisects $\angle$ ABC

    Proof: $\overline{DE} \perp \overline{BA}$ and $\overline{DF} \perp \overline{BC}$, so _____ and _____.
    are right triangles. By hypothesis, _____ $\cong$ _____.
    By identity, _____. Then $\triangle$ DEB $\cong$ $\triangle$ DFB by HL.
    Then $\angle$ ___ $\cong$ $\angle$ _____ by CPCTC. Then
    _____ by definition of angle-bisector.

Another Important Theorem

7. Complete: The locus of points in a plane that are equidistant from the endpoints of
   A line segment is the _____ of the line segment.

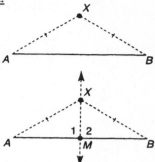

8. Fill in the blanks for the 2-part proof of the theorem in Exercise 7.
   i)   If a point is equidistant from the endpoints of a line segment,
          then it lies on the perpendicular-bisector of the line segment.

   Given: $\overline{AB}$ and point X so that AX = BX

   Prove: X lies on the perpendicular-bisector of $\overline{AB}$

   Proof: Let M represent the midpoint of $\overline{AB}$, so that
   $\overline{AM} \cong$ _____. With AX = BX, we know that $\overline{AX} \cong$ _____.
   Because _____ $\cong$ _____ by identity, it follows by SSS
   that $\triangle AMX \cong$ _____. By CPCTC, $\angle 1 \cong$ _____.
   It follows that _____ $\perp$ _____. By definition, $\overline{MX}$ is the
   perpendicular-bisector of _____ so it follows that
   _____.

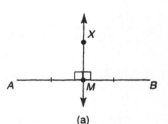

   ii)   If a point is on the perpendicular-bisector of a line segment,
          then the point is equidistant from the endpoints of the line segment.

   Given: Point X lies on the perpendicular-bisector of $\overline{AB}$.
   Prove: X is equidistant from A and B (AX = BX)

   Proof: Point X lies on the perpendicular-bisector of $\overline{AB}$.
   Then $\angle$ s 1 and 2 are congruent _____ angles.
   Also, $\overline{AM} \cong$ _____. With $\overline{XM} \cong \overline{XM}$ by _____,
   It follows that $\triangle AMX \cong$ _____ by SAS. In turn,
   $\overline{XA} \cong \overline{XB}$ by _____. Then X is equidistant from
   from A and B since AX = BX.

   (a)

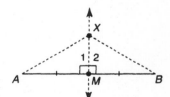

Constructions that Use the Locus Concept

9. The figure at the right represents "the locus of the vertex of

   a right triangle with hypotenuse $\overline{AB}$." Why does each angle
   marked with the square in the opening have to be a right angle?

   _____.

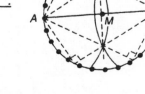

10. Construct rhombus ABCD given its diagonals $\overline{AC}$ and $\overline{BD}$.
    Note: The diagonals of a rhombus are perpendicular-bisectors
    of each other.

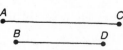

Concurrent Lines

1. Complete: A number of lines are _____ if they have exactly one point
   in common.

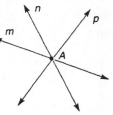 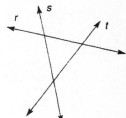

2. In the figures shown, which lines (m, n, and p *or*
   r, s, and t) are concurrent?_____.

Concurrence of Angle-Bisectors in a Triangle

3. Complete: The three angle-bisectors of a triangle are _____.

4. In the figure, the three angle-bisectors of △ABC are concurrent
   at point _____, known as the _____ of △ABC.

5. Based upon the locus theorem, "The locus of points equidistant
   from the sides of an angle is the bisector of the angle," we know
   that point E is _____ from the 3 sides of △ABC.

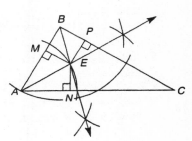

6. Just as $\overline{EN}$, $\overline{EP}$, and $\overline{EM}$ are congruent in △ABC, it is
   always possible to _____ a circle in a triangle.

7. In order to locate the incenter of a triangle, how many of the
   angle-bisectors of the triangle must be constructed? _____.

Concurrence of the Perpendicular-Bisectors of Sides of a Triangle

8. Complete: The three perpendicular-bisectors of the sides of a triangle are _____.

9. In the figure, the three perpendicular-bisectors of the sides of △ABC
   are concurrent at point _____, known as the _____ of △ABC.

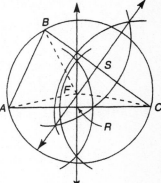

10. Based upon the locus theorem, "The locus of points equidistant
    from the endpoints of a line segment is the perpendicular-bisector
    of the line segment," we know that point F is equidistant from the
    three _____ of △ABC.

11. Just as $\overline{AF}$, $\overline{BF}$, and $\overline{CF}$ are congruent in △ABC, it is
    always possible to _____ a circle about a triangle.

12. In order to locate the circumcenter of a triangle, how many of the
    perpendicular-bisectors of sides of the triangle must be constructed? _____.

Concurrence of the Altitudes of a Triangle

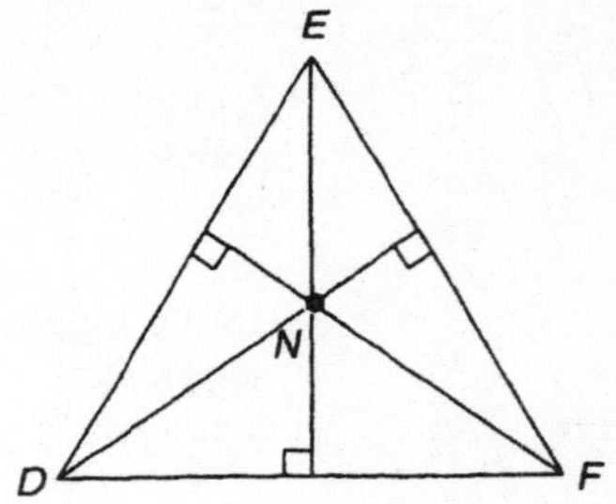

13. Complete: The three altitudes of a triangle are _____

14. In the figure, the three altitudes of △ DEF are concurrent at
     point _____, known as the _____ of △ DEF.

15. For an obtuse triangle, the orthocenter lies in the _____.
     of the triangle.

16. In order to locate the orthocenter of a triangle, how many of the
     altitudes of the triangle must be constructed? _____.

Concurrence of the Medians of a Triangle

17. Complete: The three medians of a triangle are _____ at a
     a point that is _____ of the distance from a vertex to the
     midpoint of the opposite side.

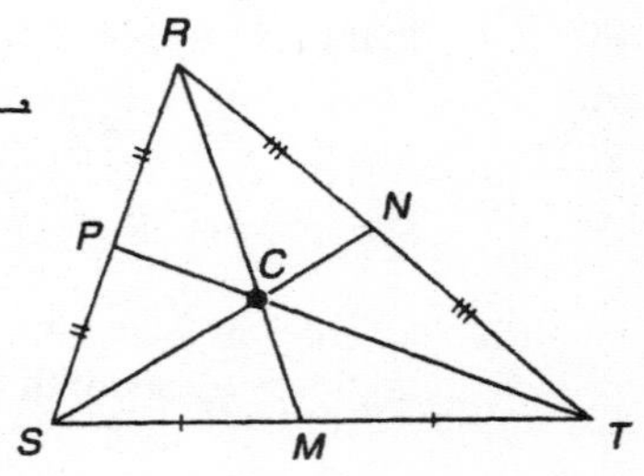

18. In the figure, the three medians of △ RST are concurrent at point _____,
     known as the _____ of △ RST.

19. In order to locate the centroid of a triangle, how many of the
     medians of the triangle must be constructed? _____.

For Exercises 20 and 21, use the figure for Exercise 18.

20. If the length of median $\overline{RM}$ is 12 inches, find:
     a) RC _____ b) CM _____.

21. If the length of $\overline{TC}$ is 10, find:
     a) TP _____ b) CP _____.

22. In isosceles △ RST, RS = RT = 15, and ST = 18.
     Medians $\overline{RZ}$, $\overline{TX}$, and $\overline{SY}$ are concurrent at
     centroid Q. Find:
     a) SZ ____. b) RZ ____ (Use the Pythagorean Theorem)

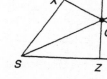

     c) RQ ____   d) QZ ____   e) SQ ____.

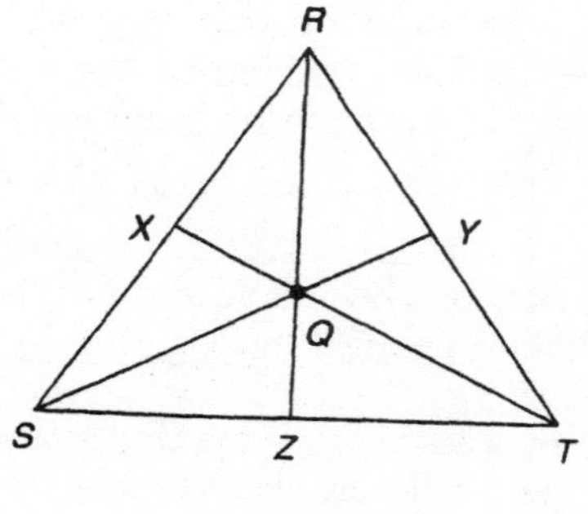

- 68 -

Initial Area Concepts

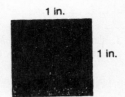
1 in.

1 in.

1. While linear units like inches or centimeters are used to measure
   length or distance, the unit of area shown measures one _____ inch.
   For convenience, we represent 1 square _____ by the symbol in^2.

2. Complete (Area Postulate): Corresponding to every bounded region is a unique
   positive number A known as the _____ of the region.

3. Complete: If two closed plane figures are congruent, then their areas are _____.

4. Complete (Area-Addition Postulate): Let R and S be two regions
   that do not overlap. Then A$_{R \cup S}$ = _____ + _____.

5. If the area of region R is 13 cm^2 and the area of region S is
   6 cm^2, find the area of the complete shaded region.   A$_{R \cup S}$ = _____.

The Area of a Rectangle

6. By counting the number of square centimeters, find the area
   of rectangle MNPQ. _____.

7. In rectangle MNPQ, the base (or length) measures b = 4 cm while the
   altitude (or width) measures h = 3 cm. Use the formula A = bh (which is
   equivalent to A = ℓ w) to find the area of rectangle MNPQ. _____.

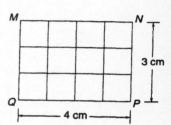
3 cm

4 cm

8. Find the area (in square inches) of a rectangle that has length ℓ = 1 foot and
   width w = 5 inches. _____.

9. How many square inches are in one square foot? _____.

10. If each side of a square has length s, then the formula for its area is _____.

The Area of a Parallelogram

11. In the figure provided, it is shown that the area of
    a parallelogram can be found by the formula A = _____.

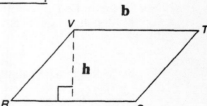
b

h

12. Complete: The area A of a parallelogram with a base of length b
    and with _____altitude of length h is given by A = bh.

13. In $\square$ MNPQ, what is the length of the altitude that corresponds:
    a) to base $\overline{MN}$? _____ b) to base $\overline{NP}$? _____.

14. Find the area of $\square$ MNPQ. _____.

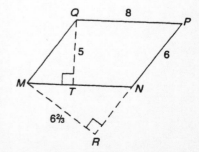
8

5

6

6⅔

The Area of a Triangle

15. Complete: The area A of a triangle whose base has length b and whose corresponding
altitude has length h is given by A = _____ .

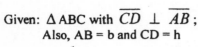

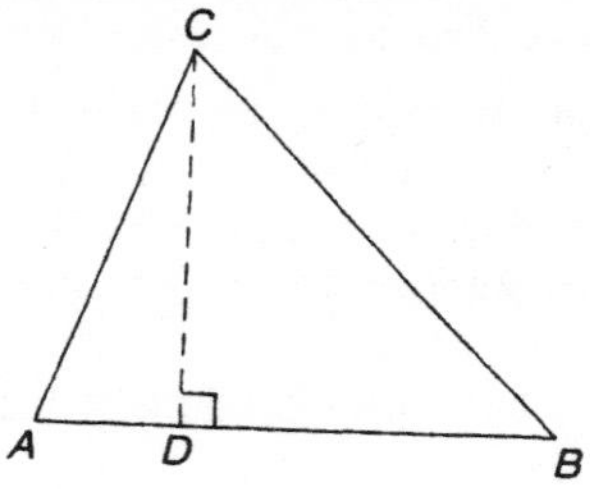

16. Fill the blanks in the proof of the theorem in Exercise 15.

Given: $\triangle$ ABC with $\overline{CD} \perp \overline{AB}$;
Also, AB = b and CD = h

Prove: A = $\frac{1}{2}$ bh.

Proof: In $\triangle$ ABC, we construct $\overline{BX} \parallel \overline{AC}$ and

$\overline{CX} \parallel$ _____. By construction, ABXC is a _____.

Then $\overline{CB}$ is a diagonal of $\square$ ABXC and $\triangle$ ABC $\cong$ _____ .

With A $_{ABC}$ = A $_{XCB}$, it follows that A $_{ABXC}$ = A $_{ABC}$ + A $_{ABC}$ OR

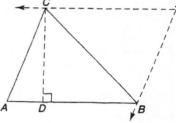

A $_{ABXC}$ = 2 · _____ . Multiplying by $\frac{1}{2}$ or dividing by 2,

we have A $_{ABC}$ = $\frac{1}{2}$ · A $_{ABXC}$ so the area of the triangle is A = _____ .

17. Find the area of $\triangle$ ABC. _____ .

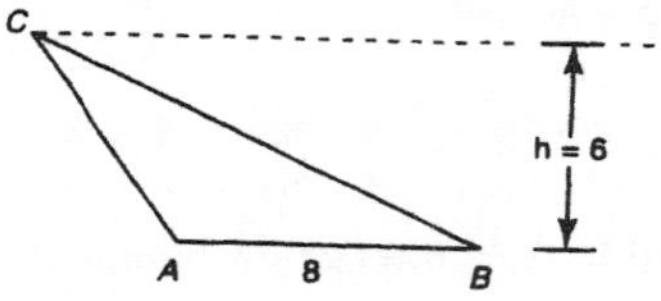

18. Complete: The area A of a right triangle whose legs have
lengths a and b is given by A = _____ .

19. Find the area of $\triangle$ PMN if PM = 8 cm and PN = 6 cm. _____ .

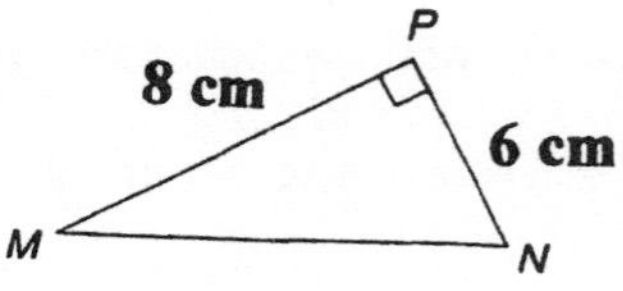

8 cm   **6 cm**

20. In the figure for Exercise 19, find:
a) MN, the length of the hypotenuse. _____ .

b) the length of the altitude to $\overline{MN}$.

Perimeter of Polygons

1. Complete: The _____ of a polygon is the sum of lengths of its sides.

2. Complete: When the area of a polygon is measured in square meters, we measure its perimeter in _____.

3. Complete formulas for:
   a) an equilateral triangle with sides of length s. $P =$ _____.
   b) a rectangle with base length b and altitude length h. $P =$ _____.

4. Find the perimeter of:
   a) an isosceles triangle with a leg of length 5 cm and base of length 6 cm. _____.

   b) rhombus ABCD with diagonals of lengths $d_1 = 6$ ft and $d_2 = 8$ ft. _____.

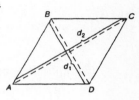

Heron's Formula

5. Complete: For a triangle with sides of lengths a, b, and c, the semiperimeter s of the triangle is given by the formula $s =$ _____.

6. Complete(Heron's Formula): For a triangle with sides of lengths a, b, and c and with semiperimeter s, the _____ of the triangle is given by the formula $A = \sqrt{s(s-a)(s-b)(s-c)}$.

7. Use Heron's Formula to find the area of a triangle with sides of lengths a = 3, b = 4, and c = 5. _____.

8. Use Heron's Formula to find the area of a triangle with sides of lengths a = 5, b = 6, and c = 7. _____.

The Area of a Trapezoid

9. Complete: By drawing an auxiliary line (diagonal), the area of a trapezoid can be shown to be equal to the sum of the areas of two _____.

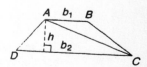

10. Complete: The area of a trapezoid whose bases have lengths b_1 and b_2 and whose altitude has length h is given by $A =$ _____.

11. Find the area of trapezoid ABCD if $b_1 = 5$ in., $b_2 = 13$ in, and h = 4 in. _____.

12. In trapezoid RSTV, RS = 6 cm and VT = 14 cm. If the area of trap. RSTV is 50 cm^2, find the length of altitude $\overline{RW}$. _____.

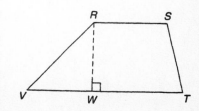

Quadrilaterals with Perpendicular Diagonals

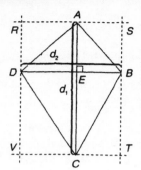

13. In the figure, quadrilateral ABCD has perpendicular diagonals
     of lengths d_1 and d_2. With auxiliary lines drawn parallel to the
     diagonals, quadrilateral RSTV is a _____ .

14. Complete: The area of any quadrilateral with perpendicular
     diagonals of lengths d_1 and d_2 is given by $\underline{A =}$ _____ .

15. Two specific types of quadrilaterals that have perpendicular
     diagonals are the kite and the _____ .

16. Find the area of rhombus MNPQ if QN = 8 ft and PM = 6 ft. _____ .

17. In the drawing for Exercise 16, suppose that QN = 24 inches and
     that the area of the rhombus is 240 in ^{2} . Find the length of the
     remaining diagonal. _____

Areas of Similar Polygons

18. In the figure shown, the lengths of the sides of the larger triangle
     are twice the lengths of the corresponding sides of the smaller triangle;

     that is, $\dfrac{s_2}{s_1} = \dfrac{2}{1}$; however, the ratio of the areas is $\dfrac{A_2}{A_1} =$ _____ .

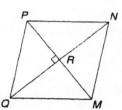
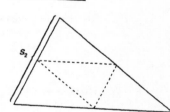

19. Complete: The ratio of the areas of two similar triangles equals the square
     of the ratio of the lengths of any two _____ sides;

     that is, $\dfrac{A_2}{A_1} =$ _____ .

20. Find the ratio of the areas of two similar triangles if a_1 = 6 inches is the
     length of a side of the smaller triangle while a_2 = 9 inches is the length of
     the corresponding side of the larger triangle. _____ .

21. The theorem stated in Exercise 19 is true for any two similar polygons.
     For the squares shown, the lengths of the sides of the smaller square are
     one-third the lengths of the corresponding sides of the larger square;

     that is, $\dfrac{s_2}{s_1} = \dfrac{1}{3}$; however, the ratio of the areas is $\dfrac{A_2}{A_1} =$ _____ .

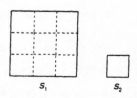

Polygons with Inscribed Circles

1. Because the angle-bisectors of a triangle are concurrent at a point (the incenter) that is equidistant from the sides of the triangle, it is always possible to _____ a circle in the triangle.

2. Does the following quadrilateral have concurrent angle-bisectors and, in turn, an inscribed circle?
   a) any rectangle? _____   b) any rhombus? _____ .
   c) any square? _____ .   d) any isosceles trapezoid? _____ .

3. Which of these regular polygons have concurrent angle-bisectors and, in turn, an inscribed circle?
   _____ .

Polygons with Circumscribed Circles

4. Because the perpendicular-bisectors of the sides a triangle are concurrent at a point (the circumcenter) that is equidistant from the vertices of the triangle, it is always possible to _____ a circle about the triangle.

5. Does the following quadrilateral have concurrent perpendicular-bisectors of sides and, in turn, a circumscribed circle?
   a) any rectangle? _____   b) any rhombus? _____ .
   c) any square? _____ .   d) any isosceles trapezoid? _____ .

6. Which of these regular polygons have concurrent perpendicular-bisectors of sides and, in turn, a circumscribed circle? _____ .

Angles in Regular Polygons

7. Use the formula $I = \dfrac{(n-2)180^0}{n}$ to find the measure of an interior angle of a regular pentagon. ___.

8. Use the formula $E = \dfrac{360^0}{n}$ to find the measure of each exterior angle of a regular octagon. _____ .

More about Regular Polygons

9. Complete: A circle can be circumscribed about or inscribed in any _____ polygon.

10. For a regular polygon, how do the center of the inscribed circle and the center of the circumscribed circle compare? _____ .

11. Complete (Definition): The _____ of a regular polygon is the common center for its inscribed and circumscribed circles.

12. Complete (Definition): A(n) _____ of a regular polygon is any line segment joining its center to one of the vertices.

13. Complete: Any radius of a regular polygon _____ the angle to which it is drawn.

14. How many radii does a regular polygon with 10 sides have? _____.

15. Find the length of a radius of a square that has sides of length 8 cm each. _____.

16. Complete (Definition): A(n) _____ of a regular polygon is any line segment drawn from the center of the regular polygon perpendicular to one of its sides.

17. Complete: Any apothem of a regular polygon _____ the side to which it is drawn.

18. Find the length of an apothem of a regular hexagon whose sides are each 6 ft in length. _____.

The Area of a Regular Polygon

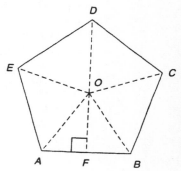

19. When the five radii of a regular pentagon are drawn, the pentagon is separated into five congruent triangles. With each side of measure s and the apothem of length a, the area of the pentagon is

$$A = \frac{1}{2}as + \frac{1}{2}as + \frac{1}{2}as + \frac{1}{2}as + \frac{1}{2}as \quad \text{or} \quad A = \frac{1}{2}a(s + s + s + s + s + s)$$

or A = _____, where P is the perimeter of the regular polygon.

20. Use the formula $A = \frac{1}{2}aP$ to find the area of a square if each side has length 6 inches. _____.

[Hint: The length of apothem must be one-half the length of a side.]

21 Use the formula $A = \frac{1}{2}aP$ to find the area of a hexagon with sides of length 10 cm each. _____.

22. Use the formula $A = \frac{1}{2}aP$ to find the area of a regular pentagon that has each side of length 9 cm and with the length of apothem being 6.2 cm. _____.

The Value of π

1. Complete: The ratio of the circumference of a circle to its length of diameter is a unique

   _____ number. For this ratio $\dfrac{C}{d}$, we use the symbol _____.

2. Some approximations for the value of π are the decimal 3.1416 and the common fraction _____.

The Circumference of a Circle

3. From the fact that $\dfrac{C}{d} = \pi$, we find the formulas for circumference to be C = _____ or C = 2π r.

4. For the circle shown, suppose that r = 7 inches. Find:
   a) the exact circumference (leave π in the answer) _____.

   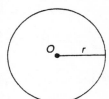

   b) the approximate circumference (replace π by $\dfrac{22}{7}$). _____.

5. The circumference of a circle is approximately 44 cm.
   To the nearest whole number, find the length of its:
   a) radius _____   b) diameter _____.

The Length of an Arc

6. While the symbol m $\overset{\frown}{AB}$ represents the degree measure of an arc, the symbol $\ell\ \overset{\frown}{AB}$ represents
   the _____ of an arc as measured in units such as inches or centimeters.

7. Because m $\overset{\frown}{AB}$ = 90^{0}, the length of $\overset{\frown}{AB}$ is what fractional
   the part of the circumference? _____.

   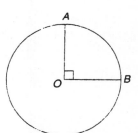

8. In $\odot$ O, the length of radius is OA = 8 cm. Find:

   a) the exact length represented by $\ell\ \overset{\frown}{AB}$. _____.

   b) the approximate length $\ell\ \overset{\frown}{AB}$ correct
       to the nearest tenth of a unit._____.

Limits

9. In some applications, there is a largest or smallest permissible value. Such numbers can be called an
   upper limit if it is largest choice possible or a lower limit if it is the _____ choice possible.

10. In a circle of radius length 5 cm, the upper limit for the
    length of chord is _____ cm.

    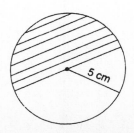

11. In ⊙O with diameter $\overline{DC}$, one side of an inscribed angle
    is chord $\overline{DA}$. With B any point between A and C, what is
    the lower limit of the measure m ∠ DAB?_____.

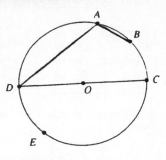

Area of a Circle

12. As the number of sides of regular polygons inscribed in a circle increases,
    the area of the regular polygon approaches that of the _____.
    Because the length of apothem a of the regular polygons approaches
    the length of radius r and the perimeter of the regular polygons approaches
    the circumference of the circle, the formula A = $\frac{1}{2}$ aP becomes

    A = _____ or, in simplified form, A = _____ for
    the area of the circle.

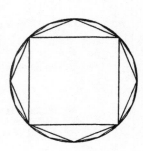

13. Complete: The area of a circle whose radius has length r is given by A = _____.

14. Where π ≈ 3.14, find the area of a circle for which the radius is r = 10 cm. _____.

15. Where π ≈ $\frac{22}{7}$, find the area of a circle whose diameter has the length d = 7 inches. _____.

16. Using the calculator value of π , find the area of the shaded semicircle
    in which OS = 3.47 meters. Give your answer correct to the nearest tenth
    of a unit. _____.

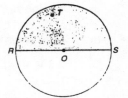

17. Using the calculator value of π , find the area of the shaded region
    bounded by concentric circles. For this ring or annulus, r = 2.3 cm
    and R = 4.2 cm. _____.

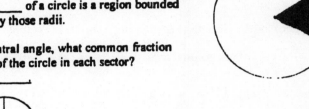

Sector of a Circle

1. Complete (definition): A _____ of a circle is a region bounded by two radii and an arc intercepted by those radii.

2. Considering the measure of the central angle, what common fraction indicates the part of the total area of the circle in each sector?

    a) _____   b) _____.

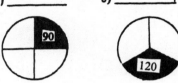

3. Complete (theorem): In a circle of radius r, the area A of a sector whose arc has measure m is given by $A =$ _____.

4. Find the approximate area of the sector shown if the radius of the circle is 8.6 inches and for which the central angle measures 110^0. Give the answer correct to the nearest tenth of a square unit. _____.

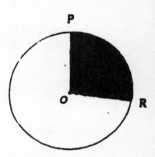

5. For the sector described in Exercise 4, the perimeter is found by P = 2r + _____.

6. Find the exact area of a sector (not shown) in a circle of radius 6 feet if the measure of the intercepted major arc is 240^0. _____.

Segment of a Circle

7. Complete (definition): A _____ of a circle is a region bounded by a chord and its minor (major) arc.

8. The plan for determining the area of a segment is given by

    $A_{segment} =$ _____ - _____.

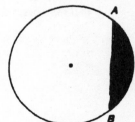

9. Find the exact area of the segment shown. _____

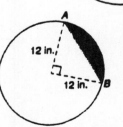

10. For the segment shown in the figure for Exercise 9, the perimeter is given by P = _____.

11. For the segment shown in Exercise 9, find its exact perimeter. _____.

Area of a Triangle with Inscribed Circle

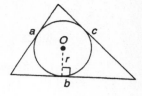

11. In the triangle shown with its inscribed circle, we draw the 3 auxiliary line segments to separate the given triangle into triangles.1, 2, and 3. Then the area of the original triangle is given by $A = A_1 + \underline{\quad} + \underline{\quad\quad}$.

12. Because the altitude of each triangle shown is r (radius of inscribed circle),

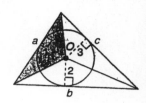

The area of the original triangle is $A = \dfrac{1}{2}ra + \dfrac{1}{2}rb + \dfrac{1}{2}rc$ or

$A = \dfrac{1}{2}r(a + b + c)$ or $\underline{A = \underline{\qquad\qquad}}$.

13. Using $A = \dfrac{1}{2}rP$, find the area of the triangle whose sides have

lengths 5 cm, 12 cm, and 13 cm and for which the radius of the inscribed circle is 2 cm. $\underline{\qquad\qquad}$.

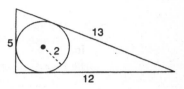

14. Heron's Formula can be used to show that the area of a triangle whose sides have lengths of 13 ft, 14 ft, and 15 ft is 84 ft^2 . Find the length of radius of the circle that can be inscribed in this triangle. $\underline{\qquad\qquad}$.

15. Find the length of the radius of the circle that can be inscribed in a right triangle with legs that measure 3 feet and 4 feet. $\underline{\qquad\qquad}$.

Prisms

1. Considering its base and whether it is right or oblique, name the type of prism shown.
    a) _____ b) _____ c) _____.

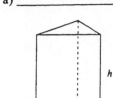 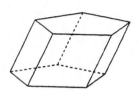 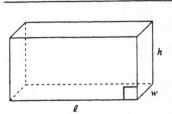

2. For the right triangular prism shown in Exercise 1a, find its number of:
    a) vertices _____   b) lateral faces _____ c) bases (base faces) _____.
    d) total faces _____   e) lateral edges _____   f) total edges _____.

3. Find the lateral area of the right triangular prism shown in Exercise 1a if the sides of each
    triangular base are 3 cm, 4 cm, and 5 cm and the altitude is h = 6 cm. _____.
    [Hint: Each triangular base is a right triangle.]

4. Find the total area of the right triangular prism described in Exercise 3. _____.

5. If the area of each of the lateral faces of the oblique pentagonal prism in Exercise 1b is 7.2 cm^2
    and the area of each base is 6.5 cm^2 , find its total area. _____.

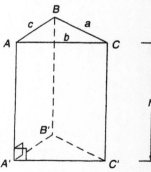

6. In the triangular prism shown, the lateral area is given by
    L = ah + bh + ch or L = h(a + b + c) or <u>L = _____</u>.

7. For the triangular prism in Exercise 6, a = 5 in, b = 6 in, and c = 4 in.
    If the lateral area is 150 in^2 , use the formula L = hP to find the length h
    of the altitude of the prism. _____.

Volume

8. The cube has dimensions for length, width, and height of 1 inch each.
    What unit of volume does it represent? _____.

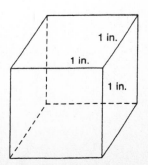

9. Complete (Volume Postulate): Corresponding to every solid is a unique positive number V known as the _____ of that solid.

10. In the figure is a right rectangular prism (box) with dimensions of 4 inches by 3 inches by 2 inches. What is its volume? _____.

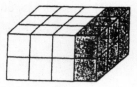

11. Complete (postulate): Where ℓ measures the length, w the width, and h the altitude of a right rectangular prism, the volume of the prism is given by V = _____.

12. Complete: The formula for the volume of a prism can be written V = Bh, where B represents the area of the _____.

13. For the regular pentagonal prism shown, the apothem of the base measures 2.1 cm while each side of the pentagon is 3 cm in length. The altitude measures 5 cm. Find:

    a) the area B of the base _____. [Recall: A = $\frac{1}{2}$ aP.]

    b) the volume V of the pentagonal prism. _____.

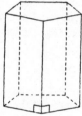

Base

14. In order to calculate area or volume of a solid figure, the dimensions of the figure must be measured in like units. Find the volume of a right rectangular prism (box) that has dimensions ℓ = 1 foot, w = 5 inches, and h = 6 inches. _____.

15. A cubic yard has dimensions of 1 yard each. Given that 1 yard = 3 feet, how many cubic feet are in a cubic yard? _____.

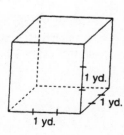

1 yd.
1 yd.
1 yd.

- 80 -

Pyramids

1. Considering its base and whether it is right or oblique, name the type of pyramid shown.
    a) _____ b) _____ c) _____.

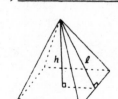

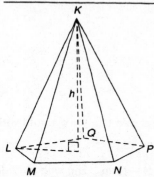

Base is a square

Base is a square

2. For the right square prism shown in Exercise 1a, find its number of:
    a) vertices _____   b) lateral faces _____   c) bases (base faces) _____.
    d) total faces _____   e) lateral edges _____   f) total edges _____.

3. For the pyramid in Exercise 1a, the length of altitude is h = 4 inches while each side
    of the square base measures s = 6 inches. Find the length of:
    a) the apothem of the square base _____ b) the slant height of the pyramid _____.

4. For the pyramid in Exercise 1a, the length of altitude is h = 8 cm and the slant height is ℓ = 10 cm.
    Find the length of:
    a) the apothem of the square base _____ b) the side of the square base _____.

5. Find the lateral area of the right pentagonal pyramid shown in Exercise 1b if each
    triangular face has an area of 30 cm^2. _____.

6. If each side of the pentagonal base measures 10 cm while the apothem measures 6.8 cm,
    find the total area of the right pentagonal pyramid described in Exercise 5. _____.

    [Hint: The area of a regular polygon is found by the formula A = $\dfrac{1}{2}$ aP.]

7. For the regular square pyramid shown, find the :
    a)   lateral area   L = _____   b) total area   T = _____ .

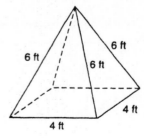

Volume

8. By experimentation, it can be shown that the pyramid with area of base B and altitude of length h has its volume given by V = _____ .

9. For a pyramid with base area B = 12 in^2 and with altitude of length h = 6 inches, find its volume.
    _____ .

10. In the figure is a right hexagonal pyramid. Each side of the base measures 4 inches while the length of altitude is 12 inches. Find:
    a) the area of the base _____ .

    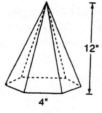
    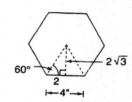

    b) the volume of the regular hexagonal pyramid. _____ .

11. For a regular square pyramid, the length of the altitude is equal to the length of each side of the square base. If the volume of the pyramid is 72 cm^3, find the length of the altitude of the pyramid.
    _____ .

Cylinders

1.   The bases of a circular cylinder are two _____ circles.

2.   If the axis (line segment joining the centers of bases) of a cylinder is perpendicular to each of its circular bases, then the cylinder is a _____ circular cylinder.

Surface Area of a Cylinder

3.   Complete: Where h is the length of its altitude and C is the circumference of each circular base, the lateral area of a right circular cylinder is given by L = hC. In terms of its length of altitude h and the length of radius r of its base, the formula can be written $\underline{L =}$ _____.

4.   For the right circular cylinder shown, find the:
     a)   exact lateral area _____.

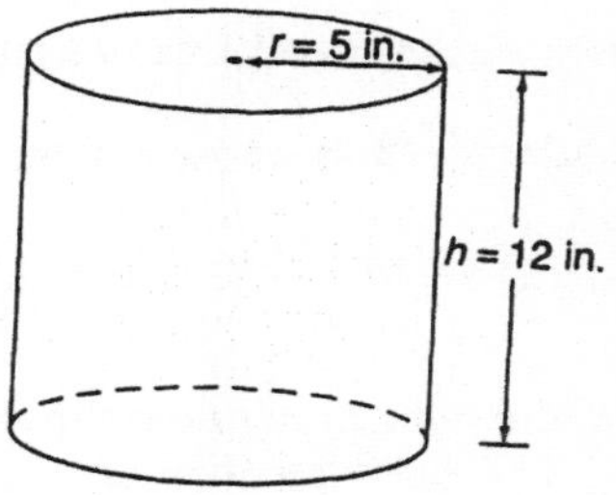

     b)   approximate lateral area _____.
          [Note: Answer to nearest tenth of a square inch.]

5. For the right circular cylinder of Exercise 4, find the:
     a)   exact total area _____.

     b)   approximate total area _____.
          [Note: Answer to nearest tenth of a square inch.])

Volume of a Cylinder

6.   The volume of a right circular cylinder with length of altitude h and area of base B is the limit of volumes (V = Bh) of inscribed regular prisms whose number of sides increases. Considering that the base of the cylinder is a circle of radius r, the formula is usually written $\underline{V =}$ _____.

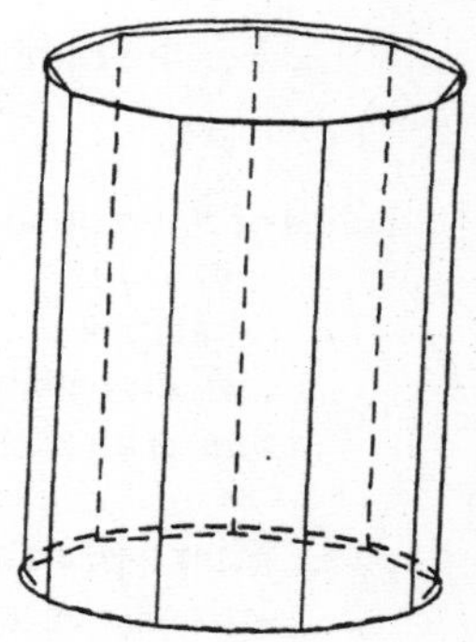

7.   For the right circular cylinder shown for Exercises 4 and 5, find the:
     a) exact volume _____.

     b) approximate volume _____.
        [Note: Answer to nearest tenth of a cubic inch.]

Cones

8. Of the two circular cones shown, the first one
    is a right circular cone while the second cone is
    a(n) _____ circular cone.

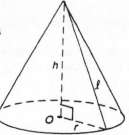

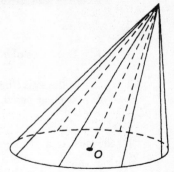

9. For the right circular cone shown in Exercise 8, find the length of the
    slant height ℓ if it is known that h = 6 cm and r = 4 cm. _____ .

Surface Area of a Right Circular Cone

10. Because the lateral area of a right circular cone is the limit of the lateral area $(L = \frac{1}{2} \ell P)$ of a

    regular pyramid whose number of sides increases, the lateral area of a right circular cone is given by

    $L = \frac{1}{2} \ell C$ or $L = \frac{1}{2} \ell (2 \pi r)$ or $\underline{L = \qquad}$.

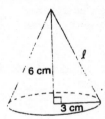

11. For the right circular cone shown, find the:
    a. exact lateral area _____ .

    b. exact total area _____ .

Volume of a Cone

12. Like the pyramid, the formula for the volume of a right circular cone with length of altitude h and

    area of base B can be written $V = \frac{1}{3} Bh$. Considering that the base of the cone is a circle of radius r,

    the formula is usually written $\underline{V = \qquad}$.

13. For the right circular cone shown in Exercise 11, find the exact volume _____ .

Solids of Revolution

14. What type of solid is generated when a rectangular region
    with dimensions of 2 ft and 5 ft is revolved about the 5 ft
    side as an axis? _____ .

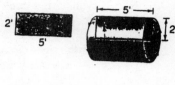

15. What type of solid is generated when a right triangular region
    with legs of lengths 4 ft and 6 ft is revolved about the 6 ft leg
    as an axis? _____ .

Polyhedrons (Polyhedra)

1.   When two planes intersect, the angle formed by two half-planes with a common edge (the line of
     Intersection) is known as a(n) _____ angle.

2.   A _____ is a solid or region of space
     bounded by (enclosed by) plane regions.

3.   Complete (Euler's equation): The number of vertices V, number of edges E, and number of faces F
     of a polyhedron are related by the equation V + F = _____.

4.   For the polyhedron shown in Exercise 2, verify Euler's equation. _____.

5.   While all prisms and pyramids are polyhedra, not all polyhedra are _____ or _____.

Regular Polyhedrons (Regular Polyhedra)

6.   For the regular polyhedra (faces are congruent polygons) shown, what are their names?
     a) _____ b)_____ c) _____ d) _____.

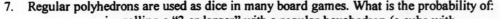

7.   Regular polyhedrons are used as dice in many board games. What is the probability of:
                  i.   rolling a "3 or larger" with a regular hexahedron (a cube with
                       faces numbered 1 through 6)? _____.

                  ii. rolling a "number larger than 3" with a regular octahedron (with
                       faces numbered 1 through 8)? _____.

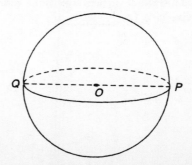

Surface Area of a Sphere

8. Complete (definition): A _____ is the set of all points in space that are located at a fixed
    distance from a fixed point(known as the center of this figure).

9. Complete: The surface area S of a sphere with length of radius r is given by S = _____.

10. Find the exact surface area of the sphere shown
     if it has a radius of length 1.2 inches. _____.

Volume of a Sphere

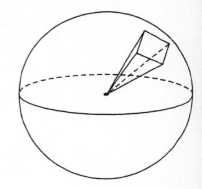

11. Suppose that a sphere is subdivided into n pyramids. Where B_1, B_2, B_3, and so on, are the areas of the bases of the pyramids, the volume V of the sphere is the limit of $\frac{1}{3}B_1 h + \frac{1}{3}B_2 h + \frac{1}{3}B_3 h + \ldots + \frac{1}{3}B_n h =$

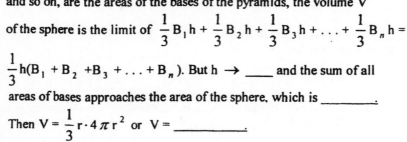

$\frac{1}{3}h(B_1 + B_2 + B_3 + \ldots + B_n)$. But h $\rightarrow$ _____ and the sum of all

areas of bases approaches the area of the sphere, which is _____.

Then $V = \frac{1}{3}r \cdot 4\pi r^2$ or $V =$ _____.

12. Use the formula $V = \frac{4}{3}\pi r^3$ to find the approximate volume of a sphere with

a radius of length 3 inches. _____.
[Hint: Use the calculator value of π and
round answer to nearest tenth of a cubic inch.]

13. A spherical tank is to be designed so that it has a volume of 200 ft^3. To the nearest tenth of a foot, find the length of radius of the sphere. _____.

More Solids of Revolution

14. What type of geometric solid is formed when a region bounded by a semicircle and its diameter of length 12 cm is revolved about the diameter? _____.

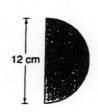

12 cm

15. Find the exact surface area of the sphere that would be generated in Exercise 14. _____.

16. Find the exact surface area of the sphere that would be generated in Exercise 14. _____.

The Rectangular Coordinate System

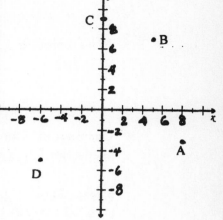

1. For the point G(-5,4), which number is the:
    a) x-coordinate? _____   b) y-coordinate? _____.

2. For each point shown, give the coordinates in the form (x,y):
    a) A _____   b) B _____   c) C _____   d) D _____.

3. Name the quadrant or axis which characterizes the location of each
    point from Exercise 2.
    a) A _____   b) B _____   c) C _____   d) D _____.

4. In the same rectangular coordinate system shown for Exercises 2 and 3,
    plot each point: E(5,-6) and F(-3,0).

Distance in the Rectangular Coordinate System

5. Find the distance between the points:
    a) (-3,4) and (5,4) _____   b) (2.1,1.3) and (2.1,7.9) _____.

6. Complete (Distance Formula): The distance between two points (x_1,y_1) and (x_2,y_2)
    is given by the formula $d =$ _____.

7. Find the distance between the points:
    a) (-1,4) and (2,0) _____   b) (0,3) and (2,9) _____.

8. The equation of a line can generally be written in the form Ax + By = C. Complete the solution
    to the following problem and leave the answer in the form Ax + By = C. Find the equation that
    describes the locus of all points that are equidistant from A(5,-1) and B(-1,7). _____.

$$\sqrt{(x-5)^2+(y-(-1))^2} = \sqrt{(x-(-1))^2+(y-7)^2}$$

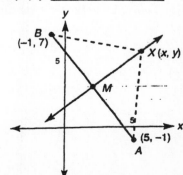

The Midpoint Formula

9. The x-coordinate of the midpoint for the line segment with endpoints (x_1,y_1) and (x_2,y_2) is found by the expression $x_1 + \dfrac{1}{2}(x_2 - x_1)$. Simplify this expression. _____ .

10. Complete (Midpoint Formula): The midpoint M of the line segment joining $A(x_1,y_1)$ and $B(x_2,y_2)$ has the form   M = (_____ , _____).

11. Find the midpoint of the line segment that joins these two points:
    a) (-1,4) and (7,0) _____   b) (a,b) and (c,d) _____ .

12. Where C is the point (-3,5), the midpoint of $\overline{CD}$ is the point M(0,-4). Find the coordinates of point D. _____ .

Symmetry in the Rectangular Coordinate System

13. Points A and B have symmetry with respect to the *y-axis*. Find B if :
    a) A = (2,-5) _____   b) A = (-3,5) _____ .

14. Points A and B have symmetry with respect to the x-*axis*. Find B if :
    a) A = (2,-5) _____   b) A = (-3,5) _____ .

15. Points A and B have symmetry with respect to the *origin*. Find B if :
    a) A = (2,-5) _____   b) A = (-3,5) _____ .

The Graph of an Equation

1. The graph of an equation is the set of all points (x,y) in the rectangular coordinate system whose ordered pairs satisfy the _____.

2. The equation $Ax + By = C$ is called a linear equation because its graph is a _____.

3. Complete the points on the graph of the linear equation $2x - 3y = 12$.
   a) where $x = 3$ (3, ___)   b) where $y = -2$ (___,-2).

Intercepts

4. For a given equation, the x-intercept (when it exists) is a point on the graph where $y =$ ____ .

5. Find the intercepts for the graph of the linear equation $3x + 4y = 24$.
   a) x-intercept (___ , ___)   b) y-intercept (___ , ___)

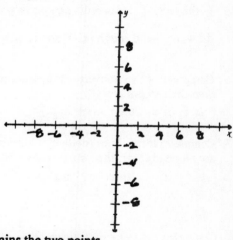

6. Using the intercepts from Exercise 5, sketch the graph of the linear equation $3x + 4y = 24$ in the rectangular coordinate system provided.

7. Because the graph of $y = 2$ is a horizontal line, it has only a(n) ___ - intercept.

8. Because the graph of $x = -5$ is a vertical line, it has only a(n) ___ - intercept.

The Slope of a Line

9. Complete (Slope Formula): The slope of the line that contains the two points (x_1,y_1) and (x_2,y_2) is given by the formula $m =$ _____.

10. Find the slope of the line that contains the points:
    a) $(-1,-4)$ and $(2,6)$ _____   b) $(0,3)$ and $(6,0)$ _____.

11. Find the slope of the line shown at the right. $m =$ _____

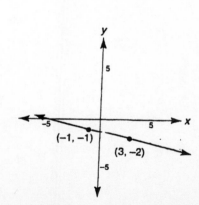

12. Find the slope of the line that contains the points:
  a) (a,0) and (0,b) _____   b) (c,d) and (e,f) _____ .

Theorems Involving the Slope of a Line

13. Given three (or more) points on a line, the slope is NOT changed when:
  a) The order of the two selected points is _____ .
  b) Two different points are _____ .

14. Are the points A(2,-3), B(5,1), and C(-4,-11) collinear? _____ .

15. In the rectangular coordinate system provided,

  draw the line through (1,-3) so that slope m = $\dfrac{1}{4}$.

16. Complete: If two nonvertical lines are parallel,
  then their slopes are _____ .
  [If $\ell_1 \parallel \ell_2$, then $m_1 = m_2$.]

17. Complete: If two lines (neither is horizontal nor vertical)
  are perpendicular, then the product of their slopes is _____ .
  [If $\ell_1 \perp \ell_2$, then $m_1 \cdot m_2 = -1$.]

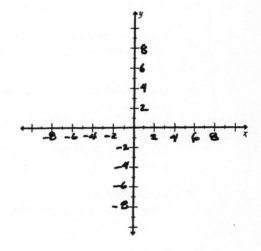

18. Given that the slope of ℓ_1 is $m_1 = \dfrac{1}{3}$, find the slope m_2 if ℓ_1 and ℓ_2 are:

  a) parallel _____   b) perpendicular _____ .

19. Given vertices A(0,0), B(a,0), C(a + b,c), and D(b,c),
  we can use slopes to show that ABCD is a _____ .

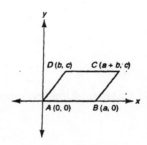

- 90 -

Formulas Used in Analytic Proof

1. Complete (Distance Formula): The distance between the points (x_1, y_1) and (x_2, y_2) in the rectangular coordinate system is given by $d =$ _____.

2. Complete (Midpoint Formula): The midpoint of the line segment with endpoints (x_1, y_1) and (x_2, y_2) $M =$ _____.

3. Complete (Slope Formula): Where $x_1 \neq x_2$, the slope of the line containing the points (x_1, y_1) and (x_2, y_2) is given by $m =$ _____.

4. Complete: Given that the slopes of lines ℓ_1 and ℓ_2 are m_1 and m_2 respectively, the lines are parallel if _____.

5. Complete: Given that the slopes of lines ℓ_1 and ℓ_2 are m_1 and m_2 respectively, the lines are perpendicular if _____. Note that neither slope is undefined or 0.

6. Which formula(s)/relationship(s) would you use to show that:
    a) two lines are parallel? _____.
    b) two line segments are congruent? _____.
    c) two line segments have the same midpoint? _____.

Using Formulas/Relationships

7. Given the points A(2a,0) and B(0,2b), find:
    a) The length of $\overline{AB}$ _____.

    b) The midpoint of $\overline{AB}$ _____.

    c) The slope of $\overline{AB}$ _____.

8. Given that the slope of line ℓ_1 is $m_1 = \dfrac{c}{d}$, find the slope of :
    a) line ℓ_2 if $\ell_1 \parallel \ell_2$. $m_2 =$ _____.

    b) line ℓ_2 if $\ell_1 \perp \ell_2$. $m_2 =$ _____.

9. If the point (a,b) lies on the graph of Ax + By = C, what equation must be true? _____.

Making a Drawing for an Analytic Proof

10. In a proof, the _____ for a point must be general, like (a,b) but not (2,3).

11. The figure drawn must satisfy the _____ of the theorem without providing further qualities. If the theorem calls for a rectangle, do not draw a square.

12. Use as many _____ coordinates for vertices of the figure as possible.

13. It is convenient to use vertical and horizontal lines because their _____ and perpendicular relationships are known.

14. Which drawing, (a) or (b), is better for a theorem of the form, "In a right triangle, ..."? _____.

15. How would you write the coordinates of the vertices of the right triangle in Exercise 14 (b) if the theorem involves the midpoint of the hypotenuse?
A = (___ , ___) , B = (___ , ___) , and C = (___ , ___) .

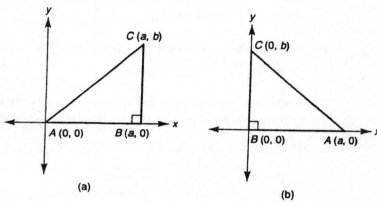

(a)

(b)

Applications Involving Drawings

16. Given ▱ MNPQ, find the coordinates of point P in terms of a, c, and d.   P = (_____ , _____)

17. In ▱ MNPQ for Exercise 16, P = (a + c,d). What expression represents the slope of:

a) both $\overline{MN}$ and $\overline{QP}$? _____.

b) both $\overline{MQ}$ and $\overline{NP}$? _____.

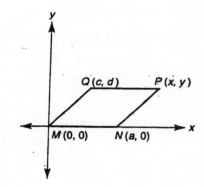

18. In ▱ ABCD, it is also known that ABCD is a rhombus, what equation follows form the fact that:

a) AB = AD? _____.

b) $\overline{AC} \perp \overline{DB}$? _____.
[Note: Diagonals not shown.]

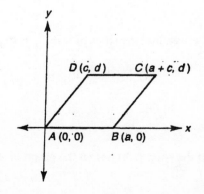

Figures Used in Analytic Proof

1. In the figure at the right, how are the pairs of opposite sides of the quadrilateral related? _____ .

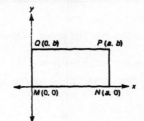

2. What type of quadrilateral is represented in the drawing shown for Exercise 1? _____ .

3. In the figure at the right, how are the sides $\overline{RT}$ and $\overline{ST}$ of the triangle related? _____ .

4. What type of triangle is shown in the drawing for Exercise 3? _____ .

5. Suppose that trapezoid ABCD is isosceles. Complete the missing coordinates of the vertices:
   A = (____ , ____), B = (a, ____), and C = (____ , ____).

Proof of Theorem 9.4.1

6. Complete: The line segment determined by the midpoints of two sides of a triangle is _____ to the third side of the triangle.

7. Complete the analytic proof of the theorem stated in Exercise 5.

   Proof: As shown in the figure, the vertices of $\triangle ABC$ are A(0,0), B(2a,0), and C(2b,2c). With M the midpoint of $\overline{CB}$, its coordinates are M = (____ , ____); likewise, with N as the midpoint of $\overline{AB}$, its coordinates are N = (____ , ____).
   Now the slope of $\overline{AC}$ is $m_{\overline{AC}}$ = _____ while the

   slope of $\overline{MN}$ is $m_{\overline{MN}}$ = _____ . Because $m_{\overline{AC}} = m_{\overline{MN}}$, we know that _____ .

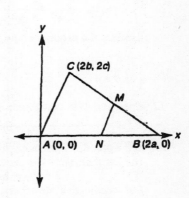

Proof of Theorem 9.4.2

8. Complete: The diagonals of a parallelogram _____ each other.

9. Complete the analytic proof of the theorem stated in Exercise 8.

Proof: As shown in the figure, the vertices of ▱ ABCD are A(0,0),
B(2a,0), C(2a + 2b,2c), and D(2b,2c). With M $_{\overline{AC}}$ the midpoint of $\overline{AC}$,

its coordinates are M $_{\overline{AC}}$ = (_____ , _____). Also, the midpoint of $\overline{DB}$

has coordinates M $_{\overline{DB}}$ = (_____ , _____). Then (a + b,c) is the common

midpoint of the two diagonals and must be the point of _____ of
the two diagonals. Thus, the diagonals must _____ each other
at point P.

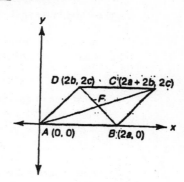

Proof of Theorem 9.4.3

10. Complete: The diagonals of a rhombus are _____ .

11. Complete the analytic proof of the theorem stated in Exercise 10.

Proof: As shown in the figure, the vertices of quadrilateral ABCD are
those of a parallelogram. Because ABCD is a rhombus, AB = AD; by
applying the Distance Formula, we see that a = _____ .
Squaring leads to a ^{2} = b ^{2} + c ^{2} (*).
To show that that the diagonals are perpendicular, we calculate their slopes:

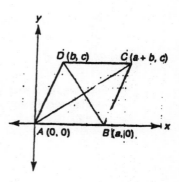

   m $_{\overline{AC}}$ = _____ and m $_{DB}$ = _____ .

Now the product of slopes is

$$m_{\overline{AC}} \cdot m_{\overline{DB}} = \frac{c}{a+b} \cdot \frac{-c}{a-b} = \frac{-c^2}{a^2-b^2}.$$

From the starred equation, we see that

$$m_{\overline{AC}} \cdot m_{\overline{DB}} = \frac{-c^2}{(b^2+c^2)-b^2} = \underline{\qquad} \text{ or } \underline{\qquad}.$$

Because the product of the slopes is –1, we know that _____ .

Theorem 9.4.4

12. Complete: If the diagonals of a parallelogram are equal in length, then the parallelogram
is a _____ .

13. In the proof of Theorem 9.4.4, we use the figure shown.
The fact that AC = DB leads to the conclusion that
        a = 0 or b = 0.
But coordinate a cannot equal 0 because points A and B (and also
D and C) would be identical. Thus, it is necessary that _____ .
With b = 0, vertices A, B, C, and D are those of a _____ .

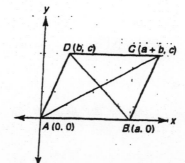

- 94 -

Linear Equations

1. Because the graphs of the equation $2x - 3y = 6$ and the more general equation $Ax + By = C$ are lines, these equations are known as _____ equations.

2. Because the graphs of the linear equations $2x + 3y = 6$ and $y = -\dfrac{2}{3}x + 2$ are the same line, these equations are known as _____ equations.

3. Are the equations $2x - 3y = 6$ and $-6x + 9y = -18$ equivalent? _____ .

The Slope-Intercept Form of a Line

4. For the graph of a linear equation in the slope-intercept form $y = mx + b$, the graph is a line with slope ____ and y-intercept ____ .

5. For the graph the equation $y = 3x - 5$, what is:
    a)   the slope? $m =$ _____   b) the y-intercept? $b =$ _____ .

6. Change (transform) the equation $2x + 3y = 5$ into the slope-intercept form. $y =$ _____ .

7. Draw the graph of the equation $y = \dfrac{2}{3}x - 3$

    in the rectangular coordinate system provided.

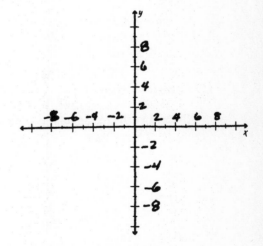

8. Find the general form $Ax + By = C$ for the line that has slope $m = -2$ and containing the point $(0,5)$. _____ .

The Point-Slope Form of a Line

9. For the graph of a linear equation in the point-slope $y - y_1 = m(x - x_1)$, the graph is a line with slope ____ and containing the point (___ , ___).

10. Find the general form $Ax + By = C$ for the line that has slope $m = \dfrac{1}{2}$ and containing

    the point $(-2,5)$. _____ .

11. Use the point-slope form to find the general form $Ax + By = C$ for the line that contains the points (-2,5) and (4,2). _____ .

12. In the form $y = mx + b$, find the equation of the line that contains the points (a,0) and (0,b). _____ .

Solving Systems of Equations

13. In a system of 2 linear equations, the solution is the _____ pair (point) at which the two graphs intersect.

14. Solve the system   { $\begin{array}{l} x + 2y = 6 \\ 2x - y = 7 \end{array}$ by using algebra. (___ , ___)

15. Solve the system   { $\begin{array}{l} x + 2y = 6 \\ 2x - y = 7 \end{array}$ by using graphing. (___ , ___)
    [Note: Use the rectangular coordinate system provided.]

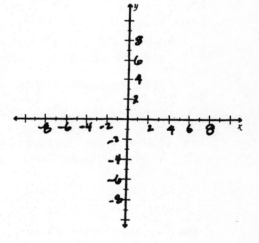

16. Solve the system   { $\begin{array}{l} x + 2y = 6 \\ 2x - y = 7 \end{array}$ by using substitution. (___ , ___)

The Sine Ratio

1. In a right triangle, the sine ratio for an angle is found by dividing the length of the leg opposite the angle by the length of the _____ .

2. In right triangle ABC with $m \angle C = 90^0$, $\sin \alpha = \dfrac{a}{c}$ while $\sin \beta = $ _____ .

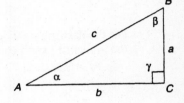

3. In the right triangle ABC shown in Exercise 2, suppose that a = 3, b = 4, and c = 5. Find:
    a) $\sin \alpha$ _____   b) $\sin \beta$ _____ .

4. In the right triangle ABC shown in Exercise 2, suppose that a = 4 and b = 5. Find:
    a) $\sin \alpha$ _____   b) $\sin \beta$ _____ .

5. The value of $\sin 27^0$ can be approximated by drawing a right triangle containing a 27^0 angle. Using a calculator, a better approximation (to 4 decimal places) would be $\sin 27^0 \approx$ _____ .

Exact Values and Behavior of the Sine Ratio

6. Using any 45^0-45^0-90^0 triangle, it can be shown that the exact value of $\sin 45^0$ is _____ .

7. Using any 30^0-60^0-90^0 triangle, it can be shown that the exact value of $\sin 30^0$ is _____ while the exact value of $\sin 60^0$ is _____ .

8. $\sin 0^0 =$ _____ and $\sin 90^0 =$ _____ .

9. When the measure θ of an acute angle increases, the value of $\sin \theta$ _____ .

10. Without using a calculator, which number is larger . . . $\sin 47^0$ or $\sin 53^0$? _____ .

Using a Calculator

11. To use a calculator to find the sine ratio of an angle measured in degrees, one must be sure that the calculator is first placed in _____ mode.

12. Using a calculator, find each number correct to 4 decimal places:
    a) sin 13^0 _____   b) sin 56.2^0 _____.

13. For the drawing provided, use the sine function to find the indicated length correct to the nearest tenth of a unit:
    a) a _____   b) b _____.

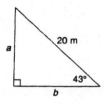

14. When found on a calculator, the notation "sin^{-1}(3/5)" is used to calculate the measure of a(n) acute _____ in a right triangle for which the ratio $\dfrac{opposite}{hypotenuse}$ equals $\dfrac{3}{5}$.

15. For the drawing provided, use the "sin^{-1} function" to find the indicated angle measure correct to the nearest degree:
    a) α _____   b) β _____.

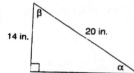

Angle of Elevation/Depression

16. With respect to the ground, a kite is flying at an angle of elevation of 67^0. If there are 72 feet of string out to the kite, what is the altitude of the kite above the ground? Answer to the nearest whole number of feet. _____.

17. On a downhill slope, a skier moves down 10 feet as she skies 120 feet across the snow. To the nearest degree, what is the angle of depression on that slope? _____.

The Cosine Ratio

1. In a right triangle, the cosine ratio for an angle is found by dividing the length of the leg
    _____ to the angle by the length of the hypotenuse.

2. In right triangle ABC with $m \angle C = 90^0$, $\sin \alpha = \dfrac{a}{c}$ while $\cos \alpha =$ _____.

3. In the right triangle ABC shown in Exercise 2, suppose that
    a = 3, b = 4, and c = 5. Find:
       a) $\cos \alpha$ _____   b) $\cos \beta$ _____.

4. In the right triangle ABC shown in Exercise 2, suppose that
    a = 5 and b = 7. Find:
       a) $\cos \alpha$ _____   b) $\cos \beta$ _____.

5. In a right triangle whose acute angles measure 27^0 and 63^0, the value of $\cos 27^0$ can be shown to
    be equal to the value of $\sin$ _____^{0}.

Exact Values and Behavior of the Sine Ratio

6. Using any 45^0-45^0-90^0 triangle, it can be shown that
    the exact value of $\cos 45^0$ is _____.

7. Using any 30^0-60^0-90^0 triangle, it can be shown that
    the exact value of $\cos 30^0$ is _____while the
    exact value of $\cos 60^0$ is _____.

8. $\cos 0^0 =$ _____ and $\cos 90^0 =$ _____.

9. When the measure θ of an acute angle increases, the value of $\cos \theta$ _____.

10. Without using a calculator, which number is larger . . . $\cos 35^0$ or $\cos 76^0$? _____.

Using a Calculator

11. Using a calculator, find each number correct to 4 decimal places:

   a) $\cos 23°$ _____   b) $\cos 62.7°$ _____.

12. For the drawing provided, use the cosine function to find the indicated length correct to the nearest tenth of a unit:

   a) c _____   b) d _____.

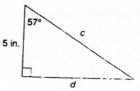

13. Which should you use, $\sin 32°$ or $\cos 32°$, if you were finding the length a in the drawing? _____.

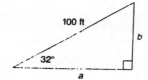

14. When found on a calculator, the notation "$\cos^{-1}(5/8)$" is used to calculate the measure

   of a(n) acute _____ in a right triangle for which the ratio $\dfrac{adjacent}{hypotenuse}$ equals $\dfrac{5}{8}$.

15. For the drawing provided, use the "$\cos^{-1}$ function" to find the indicated angle measure correct to the nearest degree:

   a) α _____   b) β _____.

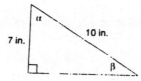

Angle of Elevation/Depression

16. At a point 200 feet from the base of a cliff, the top of the cliff is seen through an angle of elevation of $37°$. How tall is the cliff? _____.

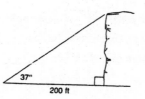

17. Use the drawing provided to find the length of the radius of a regular pentagon whose apothem measures 12 inches.

   _____.

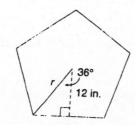

The Tangent Ratio

1. While the cosine ratio for an acute angle of a right triangle is given by $\dfrac{adjacent}{hypotenuse}$,

   the tangent of that angle is the ratio _____ .

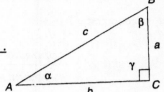

2. Where m $\angle$ C = 90^0, sin α = $\dfrac{a}{c}$, cos α = $\dfrac{b}{c}$, and tan α _____.

3. In the right triangle ABC shown in Exercise 2, suppose that
   a = 8 and b = 15. Find:
   a) tan α _____ b) tan β _____.

4. In the right triangle ABC shown in Exercise 2, suppose that
   a = 6 and c = 10. Find:
   a) tan α _____ b) tan β _____.

Exact Values and Behavior of the Sine Ratio

5. Using any 45^0-45^0-90^0 triangle, it can be shown that
   the exact value of tan 45^0 is _____.

6. Using any 30^0-60^0-90^0 triangle, it can be shown that
   the exact value of tan 30^0 is _____while the
   exact value of tan 60^0 is _____.

7. tan 0^0 = _____ while tan 90^0 is _____.

8. When the measure θ of an acute angle increases, the value of tan θ _____.

Using a Calculator

9. Using a calculator, find each number correct to 4 decimal places:
   a) tan 37^0 _____ b) tan 89.5^0 _____.

10. For the drawing provided, use the tangent function to find
    the indicated length y. _____.

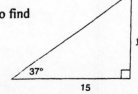

11. Use the calculator and the "tan^{-1}" function to find θ to the nearest degree. _____ .

Other Trigonometric Ratios

12. While sine $\alpha = \dfrac{opposite}{hypotenuse}$ in a right triangle, its reciprocal is cosecant $\alpha =$ _____ .

13. While cosine $\alpha = \dfrac{adjacent}{hypotenuse}$ in a right triangle, its reciprocal is secant $\alpha =$ _____ .

14. While tangent $\alpha = \dfrac{opposite}{adjacent}$ in a right triangle, its reciprocal is cotangent $\alpha =$ _____ .

15. Use the calculator to find cot 16^0 correct to 4 decimal places. _____ .
    [Hint: cot $\beta = 1 \div$ tan β.]

16. Find: a) sec α if cos $\alpha = \dfrac{2}{5}$. _____ b) csc β if sin $\beta = \dfrac{9}{13}$ _____ .

Applications

17. Which ratio would you use to find the length a . . .
    sin 52^0, cos 52^0, or tan 52^0? _____ .

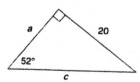

18. The top of a lookout tower is seen from a point 270 ft from its base.
    If the angle of elevation is 37^0, how tall is the tower? _____ .

The Area of a Triangle

1. Complete: The area of a triangle equals one-half the product of the lengths of two sides of a triangle and the sine of their _____ angle.

2. Fill in the blanks to complete the proof of the theorem in Exercise 1.

    Given: Acute triangle ABC

    Prove: $A = \dfrac{1}{2} bc \sin \alpha$

    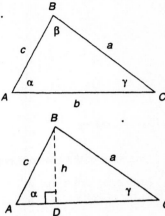

    Proof: The area of $\triangle ABC$ is given by $A = \dfrac{1}{2} bh$,

    where h is the _____ of the indicated altitude. In right triangle ABD, $\sin \alpha =$ _____;
    in turn, h = _____. Substitution for h into

    $A = \dfrac{1}{2} bh$ leads to $A = \dfrac{1}{2} b(_____)$ or $A =$ _____.

3. In addition to the formula for calculating area found in Exercise 2 are these two additional forms:

    $A = \dfrac{1}{2} ac \sin \beta$ and _____.

4. To the nearest tenth of a square meter, find the area of the triangle shown at the right. _____ .

    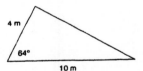

The Law of Sines

5. Because the area of a triangle is unique, we know that

    $\dfrac{1}{2} bc \sin \alpha = \dfrac{1}{2} ac \sin \beta = A = \dfrac{1}{2} ab \sin \gamma$.

    Dividing each expression by $\dfrac{1}{2} abc$, we obtain the

    Law of Sines _____ .

6. Use the Law of Sines to find length x. _____

    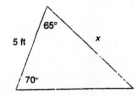

7. Use the Law of Sines to find the measure α to the nearest degree. _____ .

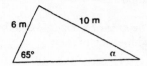

The Law of Cosines

8. One form of the Law of Cosines, namely $c^2 = a^2 + b^2 - 2ab \cos \gamma$, is proven in the textbook. Two additional forms are: $a^2 =$ _____ and
$b^2 =$ _____ .

9. Use the Law of Cosines to determine length x. _____ .

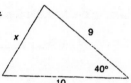

10. Using the Law of Cosines, find to the nearest degree the measure β of the indicated angle._____ .

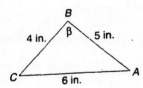

Chapter 1 Line and Angle Relationships

SECTION 1.1: Sets, Statements, and Reasoning

1. **a.** Not a statement.

    b. Statement; true

    c. Statement; true

    d. Statement; false

5. Conditional

9. Simple

13. H: The diagonals of a parallelogram are perpendicular.

    C: The parallelogram is a rhombus.

17. First, write the statement in "If, then" form. If a figure is a square, then it is a rectangle.

    H: A figure is a square.

    C: It is a rectangle.

21. True

25. Induction

29. Intuition

33. Angle 1 looks equal in measure to angle 2.

37. *A Prisoner of Society* might be nominated for an Academy Award.

41. Angles 1 and 2 are complementary.

45. None

49. Marilyn is a happy person.

53. Not valid

SECTION 1.2: Informal Geometry and Measurement

1. $AB < CD$

5. One; none

9. Yes; no; yes

13. Yes; no

17. **a.** 3

    b. $2\frac{1}{2}$

21. Congruent; congruent

25. No

29. Congruent

33. $\overline{AB}$

37. $x + x + 3 = 21$
    $\quad\quad 2x = 18$
    $\quad\quad\ \ x = 9$

41. 71

45. $32.7 \div 3 = 10.9$

49. N 22° E

SECTION 1.3: Early Definitions and Postulates

1. AC

5. $\frac{1}{2}$m · 3.28 ft/m = 1.64 feet

9. **a.** *A-C-D*

    b. A, B, C or B, C, D or A, B, D

13. **a.** m and t

    b. m and $\overrightarrow{AD}$ or $\overrightarrow{AD}$ and t

17. $2x + 1 + 3x = 6x - 4$
    $\quad\quad 5x + 3 = 6x - 4$
    $\quad\quad\ \ -1x = -7$
    $\quad\quad\quad\ \ x = 7;\ AB = 38$

21. **a.**

    b.

    c.

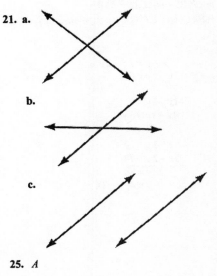

25. A

29. Given: $\overline{AB}$ and $\overline{CD}$ as shown $(AB > CD)$
Construct $\overline{MN}$ on line l so that
$MN = AB + CD$

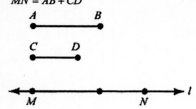

33. a. No

   b. Yes

   c. No

   d. Yes

37. Nothing

SECTION 1.4: Angles and Their Relationships

1. a. Acute

   b. Right

   c. Obtuse

5. Adjacent

9. a. Yes

   b. No

13. $m\angle FAC + m\angle CAD = 180$
$\angle FAC$ and $\angle CAD$ are supplementary.

17. $42°$

21. $\begin{aligned} x + y &= 2x - 2y \\ x + y + 2x - 2y &= 64 \end{aligned}$

$\begin{aligned} -1x + 3y &= 0 \\ 3x - 1y &= 64 \end{aligned}$

$\begin{aligned} -5x + 2y &= 10 \\ 5x + 2y &= 78 \\ \hline 22y &= 88 \\ y = 4; \ x &= 14 \end{aligned}$

25. $\begin{aligned} x + y &= 180 \\ x &= 24 + 2y \end{aligned}$

$\begin{aligned} x + y &= 180 \\ x - 2y &= 24 \end{aligned}$

$\begin{aligned} -2x + 2y &= 360 \\ x - 2y &= 24 \\ \hline 3x &= 384 \\ x = 128; \ y &= 52 \end{aligned}$

29. $\begin{aligned} x - 92 + (92 - 53) &= 90 \\ x - 92 + 39 &= 90 \\ x - 53 &= 90 \\ x &= 143 \end{aligned}$

33. Given: Obtuse $\angle MRP$
Construct: Rays RS, RT, and RU so that $\angle MRP$ is divided into 4 $\cong$ angles.

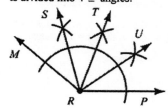

37. It appears that the two sides opposite $\angle$ s A and B are congruent.

41. $\begin{aligned} 90 + x + x &= 360 \\ 2x &= 270 \\ x &= 135 \end{aligned}$

SECTION 1.5: Introduction to Geometric Proof

1. Division Property of Equality or Multiplication Property of Equality

5. Multiplication Property of Equality

9. Angle-Addition Property

13. $\overrightarrow{EG}$ bisects $\angle DEF$

17. $2x = 10$

21. $6x - 3 = 27$

25. 1. $2(x + 3) - 7 = 11$

   2. $2x + 6 - 7 = 11$

   3. $2x - 1 = 11$

   4. $2x = 12$

   5. $x = 6$

29. 1. Given

  2. If an angle is bisected, then the two angles formed are equal in measure.

  3. Angle-Addition Postulate

  4. Substitution

  5. Distribution Property

  6. Multiplication Property of Equality

33. $5 \cdot x + 5 \cdot y = 5(x + y)$

SECTION 1.6: Relationships: Perpendicular Lines

1. 1. Given

  2. If 2 $\angle$ s are $\cong$, then they are equal in measure.

  3. Angle-Addition Postulate

  4. Addition Property of Equality

  5. Substitution

  6. If 2 $\angle$ s are = in measure, then they are $\cong$.

5. Given: Point N on line s.
Construct: Line m through N so that $m \perp s$.

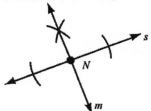

9. Given: Triangle ABC
Construct: The perpendicular bisectors of each side, $\overline{AB}$, $\overline{AC}$, and $\overline{BC}$.

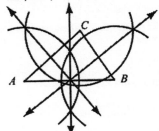

13. No; Yes; No

17. No; Yes; Yes

21. 1. $M - N - P - Q$ on $\overline{MQ}$ **1.** Given

  2. $MN + NQ = MQ$ **2.** Segment-Addition Postulate

  3. $NP + PQ = NQ$ **3.** Segment-Addition Postulate

  4. $MN + NP + PQ = MQ$ **4.** Substitution

25. In space, there are an infinite number of lines which perpendicularly bisect a given line segment at its midpoint.

SECTION 1.7: The Formal Proof of a Theorem

1. H: A line segment is bisected.

  C: Each of the equal segments has half the length of the original segment.

5. H: Each is a right angle.

  C: Two angles are congruent.

9. Given: $\overline{AB} \perp \overline{CD}$
Prove: $\angle AEC$ is a right angle.

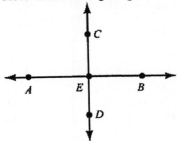

13. Given: Lines l and m
Prove: $\angle 1 \cong \angle 2$ and $\angle 3 \cong \angle 4$

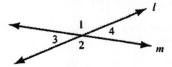

17. $m\angle 1 = m\angle 3$
  $3x + 10 = 4x - 30$
    $x = 40; \ m\angle 1 = 130°$

21. 1. Given

  2. If 2 $\angle$ s are comp., then the sum of their measures is 90.

  3. Substitution

  4. Subtraction Property of Equality

  5. If 2 $\angle$ s are = in measure, then they are $\cong$.

25. 1. Given

    2. $\angle ABC$ is a right $\angle$.

    3. The measure of a rt. $\angle = 90$.

    4. Angle-Addition Postulate

    6. $\angle 1$ is comp. to $\angle 2$.

29. The supplement of an acute angle is obtuse.
Given: $\angle 1$ is supp to $\angle 2$
      $\angle 2$ is an acute $\angle$
Prove: $\angle 1$ is an obtuse $\angle$

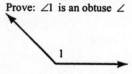

STATEMENTS	REASONS
1. $\angle 1$ is supp to $\angle 2$	1. Given
2. $m\angle 1 + m\angle 2 = 180$	2. If 2 $\angle$s are supp., the sum of their measures is 180.
3. $\angle 2$ is an acute $\angle$	3. Given
4. $m\angle 2 = x$ where $0 < x < 90$	4. The measure of an acute $\angle$ is between 0 and 90.
5. $m\angle 1 + x = 180$	5. Substitution (#4 into #3)
6. x is positive $\therefore m\angle 1 < \angle 180$	6. If $a + p_1 = b$ and p_1 is positive, then $a < b$.
7. $m\angle 1 = 180 - x$	7. Substitution Prop of Eq. (#5)
8. $-x < 0 < 90 - x$	8. Subtraction Prop of Ineq. (#4)
9. $90 - x < 90 < 180 - x$	9. Addition Prop. or Ineq. (#8)
10. $90 - x < 90 < m\angle 1$	10. Substitution (#7 into #9)
11. $90 < m\angle 1 < 180$	11. Transitive Prop. of Ineq (#6 & #10)
12. $\angle 1$ is an obtuse $\angle$	12. If the measure of an angle is between 90 and 180, then the $\angle$ is obtuse.

REVIEW EXERCISES

1. Undefined terms, defined terms, axioms or postulates, theorems.

2. Induction, deduction, intuition

3. 1. Names the term being defined.

    2. Places the term into a set or category.

    3. Distinguishes the term from other terms in the same category.

    4. Reversible

4. Intuition

5. Induction

6. Deduction

7. H: The diagonals of a trapezoid are equal in length.

    C: The trapezoid is isosceles.

8. H: The parallelogram is a rectangle.

    C: The diagonals of a parallelogram are congruent.

9. No conclusion

10. Jody Smithers has a college degree.

11. Angle A is a right angle.

12. C

13. $\angle RST$, $\angle S$, more than 90°.

14. Diagonals are $\perp$ and they bisect each other.

15.

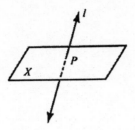

16.

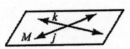

17.

18. $2x + 15 = 3x - 2$
    $17 = x$
    $x = 17;\ m\angle ABC = 98°$

19. $2x + 5 + 3x - 4 = 86$
    $5x + 1 = 86$
    $5x = 85$
    $x = 17;\ m\angle DBC = 47°$

20. $3x - 1 = 4x - 5$
    $4 = x$
    $x = 4;\ AB = 22$

21. $4x - 4 + 5x + 2 = 25$
    $9x - 2 = 25$
    $9x = 27$
    $x = 3;\ MB = 17$

22. $2 \cdot CD = BC$
    $2(2x + 5) = x + 28$
    $4x + 10 = x + 28$
    $3x = 18$
    $x = 6;\ AC = BC = 6 + 28 = 34$

23. $7x - 21 = 3x + 7$
    $4x = 28$
    $x = 7$
    $m\angle 3 = 49 - 21 = 28°$
    $\therefore m\angle FMH = 180 - 28 - 152°$

24. $4x + 1 + x + 4 = 180$
    $5x + 5 = 180$
    $5x = 175$
    $x = 35$
    $m\angle 4 = 35 + 4 = 39°$

25. **a.** Point M

    b. $\angle JMH$

    c. $\overline{MJ}$

    d. $\overleftrightarrow{KH}$

26. $2x - 6 + 3(2x - 6) = 90$
    $2x - 6 + 6x - 18 = 90$
    $8x - 24 = 90$
    $8x = 114$
    $x = 14\frac{1}{4}$

    $m\angle EFH = 3(2x - 6) = 3\left(28\frac{1}{2} - 6\right)$
    $= 3 \cdot 22\frac{1}{2}$
    $= 67\frac{1}{2}°$

27. $x + (40 + 4x) = 180$
    $5x + 40 = 180$
    $5x = 140$
    $x = 28$

28. **a.** $2x + 3 + 3x - 2 + x + 7 = 6x + 8$

    b. $6x + 8 = 32$
    $6x = 24$
    $x = 4$

    c. $2x + 3 = 2(4) + 3 = 11$
    $3x - 2 = 3(4) - 2 = 10$
    $x + 7 = 4 + 7 = 11$

29. The measure of angle 3 is less than 50.

30. The four foot board is 48 inches. Subtract 6 inches on each end leaving 36 inches.
    $4(n - 1) = 36$
    $4n - 4 = 36$
    $4n = 40$
    $n = 10$
    $\therefore$ 10 pegs will fit on the board.

31. S

32. S

33. A

34. S

35. N

36. **2.** $\angle 4 \cong \angle P$

    3. $\angle 1 \cong \angle 4$

    4. If 2 $\angle$ s are $\cong$, then their measures are $=$.

    5. Given

    6. $m\angle 2 = m\angle 3$

    7. $m\angle 1 + m\angle 2 = m\angle 4 + m\angle 3$

    8. Angle-Addition Postulate

    9. Substitution

    10. $\angle TVP \cong \angle MVP$

37. Given: $\overline{KF} \perp \overline{FH}$
          $\angle JHK$ is a right $\angle$
    Prove: $\angle KFH \cong \angle JHF$

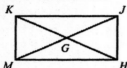

STATEMENTS	REASONS
1. $\overline{KF} \perp \overline{FH}$	**1.** Given
2. $\angle KFH$ is a right $\angle$	**2.** If 2 segments are $\perp$, then they form a right $\angle$.
3. $\angle JHF$ is a right $\angle$	**3.** Given
4. $\angle KFH \cong \angle JHF$	**4.** Any two right $\angle$s are $\cong$.

38. Given: $\overline{KH} \cong \overline{FJ}$
          G is the midpoint of both $\overline{KH}$ and $\overline{FJ}$
    Prove: $\overline{KG} \cong \overline{GJ}$

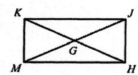

STATEMENTS	REASONS
1. $\overline{KH} \cong \overline{FJ}$ G is the midpoint of both $\overline{KH}$ and $\overline{FJ}$	**1.** Given
2. $\overline{KG} \cong \overline{GJ}$	**2.** If 2 segments are $\cong$, then their midpoints separate these segments into 4 $\cong$ segments.

39. Given: $\overline{KF} \perp \overline{FH}$
Prove: $\angle KFH$ is comp to $\angle JHF$

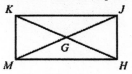

STATEMENTS	REASONS
1. $\overline{KF} \perp \overline{FH}$	1. Given
2. $\angle KFH$ is comp. to $\angle JFH$	2. If the exterior sides of 2 adjacent $\angle$s form $\perp$ rays, then these $\angle$s are comp.

40. Given: $\angle 1$ is comp. to $\angle M$
    $\angle 2$ is comp. to $\angle M$
Prove: $\angle 1 \cong \angle 2$

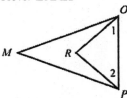

STATEMENTS	REASONS
1. $\angle 1$ is comp. to $\angle M$	1. Given
2. $\angle 2$ is comp. to $\angle M$	2. Given
3. $\angle 1 \cong \angle 2$	3. If 2 s are comp. to the same , then these angles are $\cong$.

41. Given: $\angle MOP \cong \angle MPO$
    $\overrightarrow{OR}$ bisects $\angle MOP$
    $\overrightarrow{PR}$ bisects $\angle MPO$
Prove: $\angle 1 \cong \angle 2$

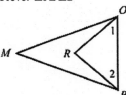

STATEMENTS	REASONS
1. $\angle MOP \cong \angle MPO$	1. Given
2. $\overrightarrow{OR}$ bisects $\angle MOP$ $\overrightarrow{PR}$ bisects $\angle MPO$	2. Given
3. $\angle 1 \cong \angle 2$	3. If 2 s are $\cong$, then their bisectors separate these $\angle$s into four $\cong$ $\angle$s.

42. Given: $\angle 4 \cong \angle 6$
   Prove: $\angle 5 \cong \angle 6$

STATEMENTS	REASONS
1. $\angle 4 \cong \angle 6$	1. Given
2. $\angle 4 \cong \angle 5$	2. If 2 angles are vertical $\angle$s then they are $\cong$.
3. $\angle 5 \cong \angle 6$	3. Transitive Prop.

43. Given: Figure as shown
   Prove: $\angle 4$ is supp. to $\angle 2$

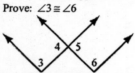

STATEMENTS	REASONS
1. Figure as shown	1. Given
2. $\angle 4$ is supp. to $\angle 2$	2. If the exterior sides of 2 adjacent $\angle$s form a line, then the $\angle$s are supp.

44. Given: $\angle 3$ is supp. to $\angle 5$
   $\angle 4$ is supp. to $\angle 6$
   Prove: $\angle 3 \cong \angle 6$

STATEMENTS	REASONS
1. $\angle 3$ is supp to $\angle 5$ $\angle 4$ is supp to $\angle 6$	1. Given
2.	If 2 lines intersect, the vertical angles 2. formed are $\cong$.
3. $\angle 3 \cong \angle 6$	3. If 2 $\angle$s are supp to congruent angles, then these angles are $\cong$.

45. Given: $\overline{VP}$
   Construct: $\overline{VW}$ such that $VW = 4 \cdot VP$

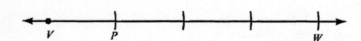

46. Construct a 135° angle.

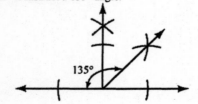

47. Given: Triangle *PQR*
   Construct: The three angle bisectors.

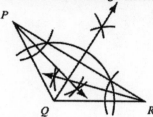

It appears that the three angle bisectors meet at one point inside the triangle.

48. Given: $\overline{AB}$, $\overline{BC}$, and $\angle B$ as shown
   Construct: Triangle *ABC*

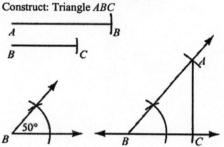

49. Given: $m\angle B = 50°$
   Construct: An angle whose measure is 20°.

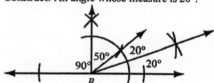

50. $m\angle 2 = 270°$

CHAPTER TEST

1. Induction

2. $\angle CBA$ or $\angle B$

3. $\overline{AP} + \overline{PB} = \overline{AB}$

4. **a.** point
   b. line

5. **a.** Right
   b. Obtuse

6. **a.** Supplementary
   b. Congruent

7. $m\angle MNP = m\angle PNQ$

8. **a.** right
   b. supplementary

9. Kianna will develop reasoning skills.

10. $3.2 + 7.2 = 10.4$ in.

11. **a.** $x + x + 5 = 27$
    $2x + 5 = 27$
    $2x = 22$
    $x = 11$

    b. $x + 5 = 11 + 5 = 16$

12. $m\angle 4 = 35°$

13. **a.** $x + 2x - 3 = 69$
    $3x - 3 = 69$
    $3x = 72$
    $x = 24$

    b. $m\angle 4 = 2(24) - 3 = 45°$

14. **a.** $m\angle 2 = 137°$
    b. $m\angle 2 = 43°$

15. **a.** $2x - 3 = 3x - 28$
    $x = 25$

    b. $m\angle 1 = 3(25) - 28 = 47°$

16. **a.** $2x - 3 + 6x - 1 = 180$
    $8x - 4 = 180$
    $8x = 184$
    $x = 23$

    b. $m\angle 2 = 6(23) - 1 = 137°$

17. $x + y = 90$

18.

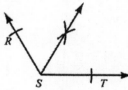

19.

20. **1.** Given

   2. Segment-Addition Postulate

   3. Segment-Addition Postulate

   4. Substitution

21. **1.** $2x - 3 = 17$

   2. $2x = 20$

   3. $x = 10$

22. **1.** Given

   2. 90°

   3. Angle-Addition Postulate

   4. 90°

   5. Given

   6. Definition of Angle-Bisector

   7. Substitution

   8. $m\angle 1 = 45°$

Chapter 2: Parallel Lines

SECTION 2.1: The Parallel Postulate and Special Angles

1. **a.** 108°

   b. 72°

5. **a.** No

   b. Yes

   c. No

9. **a.** $m\angle 3 = 87°$; $\angle 3$ is vertical to $\angle 2$.

   b. $m\angle 6 = 87°$; $\angle 6$ corresponds to $\angle 2$.

   c. $m\angle 1 = 93°$; $\angle 1$ is supplementary to $\angle 2$.

   d. $m\angle 7 = 87°$; $\angle 7$ corresponds to $\angle 3$.

13. **a.** $m\angle 2 = 68°$; $\angle 2$ is supp. to $\angle 1$.

    b. $m\angle 4 = 112°$; $\angle 4$ is vertical to $\angle 1$.

    c. $m\angle 5 = 112°$; $\angle 5$ is an alternate interior $\angle$ to $\angle 4$.

    d. $m\angle MOQ = 34°$

    $m\angle MON = m\angle 2 = 68°$

    $m\angle MOQ = \dfrac{1}{2}$ of $m\angle MON = 34°$.

17. Angles 3 and 5 are supp. because they are interior angles on the same side of the transversal. Angles 5 and 6 are also supp. This leads to a system of 2 equations with 2 variables.

    $(6x + y) + (8x + 2y) = 180$
    $(8x + 2y) + (4x + 7y) = 180$

    Simplifying yields,
    $14x + 3y = 180$
    $12x + 9y = 180$

    Dividing the 2nd equation by −3 gives
    $14x + 3y = 180$
    $-4x - 3y = -60$

    Addition gives
    $10x = 120$
    $x = 12$

    Using $14x + 3y = 180$ and $x = 12$ we get
    $14(12) + 3y = 180$
    $168 + 3y = 180$
    $3y = 12$
    $y = 4$

    $m\angle 6 = 4(12) + 7(4) = 76°$
    $\therefore m\angle 7$ also $= 76°$.

21. Given: $\overline{CE} \parallel \overline{DF}$; trans. $\overline{AB}$
    $\angle JHK$ is a right $\angle$
    Prove: $\angle KFH \cong \angle JHF$

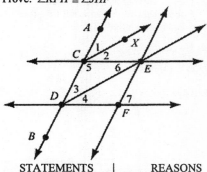

STATEMENTS	REASONS
1. $\overline{CE} \parallel \overline{DF}$; trans. $\overline{AB}$	1. Given
2. $\angle ACE \cong \angle ADF$	2. If 2 $\parallel$ lines are cut by a trans., then the corresponding $\angle$s are $\cong$.
3. $\overline{CX}$ bisects $\angle ACE$ $\overline{DE}$ bisects $\angle CDF$	3. Given
4. $\angle 1 \cong \angle 3$	4. If two $\angle$s are $\cong$, then their bisectors separate these $\angle$s into four $\cong$ $\angle$s.

25. **a.** $\angle 4 \cong \angle 2$ and $\angle 5 \cong \angle 3$

    b. 180°

    c. 180°

29. No

33. Given: Triangle MNQ with obtuse $\angle MNQ$
    Construct: $\overline{MR} \perp \overline{NQ}$
    (Hint: Extend $\overline{NQ}$)

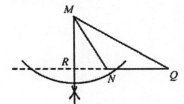

SECTION 2.2: Indirect Proof

1. If Juan wins the state lottery, then he will be rich.
   Converse: If Juan is rich, then he won the state lottery. FALSE.
   Inverse: If Juan does not win the state lottery, then he will not be rich. FALSE.
   Contrapositive: If Juan is not rich, then he did not win the state lottery. TRUE.

5. No conclusion.

9. (a) (b) (e)

13. Given: $\angle AOC \not\cong \angle AFE$
    Prove: $\overline{DC} \not\parallel \overline{EG}$

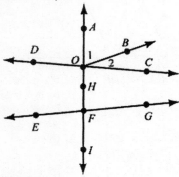

   Proof:
   Assume that $\overline{DC} \parallel \overline{EG}$. If they are $\parallel$, then
   $\angle AOC \cong \angle AFE$ because they are corresponding angles. But this contradicts the Given information. Therefore, our assumption is false and $\overline{DC} \not\parallel \overline{EG}$.

17. Assume that the angles are vertical angles. If they are vertical angles, then they are congruent. But this contradicts the hypothesis that the two angles are not congruent. Hence, our assumption must be false and the angles are not vertical angles.

21. Given: M is a midpoint of $\overline{AB}$.

   Prove: M is the only midpoint of $\overline{AB}$.
   Proof: If M is a midpoint of $\overline{AB}$, then
   $AM = \frac{1}{2} \cdot AB$. Assume that N is also a midpoint
   of $\overline{AB}$ so that $AN = \frac{1}{2} \cdot AB$. By substitution
   $AM = AN$.

   By the Segment-Addition Postulate,
   $AM = AN + NM$. Using substitution again,
   $AN + NM = AN$. Subtracting gives $NM = 0$.
   But this contradicts the Ruler Postulate which states that the measure of a line segment is a positive number. Therefore, our assumption is wrong and M is the only midpoint for $\overline{AB}$.

SECTION 2.3: Proving Lines Parallel

1. $l \parallel m$

5. $l \not\parallel m$ (Sum = 181°)

9. None

13. None

17. 1. Given

    2. If 2 $\angle$s are comp. to the same $\angle$, then they are $\cong$.

    3. $\overline{BC} \parallel \overline{DE}$

21. Given: $\overline{DE}$ bisects $\angle CDA$
$\angle 3 \cong \angle 1$
Prove: $\overline{ED} \parallel \overline{AB}$

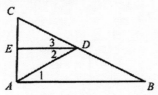

STATEMENTS	REASONS
1. $\overline{CE} \parallel \overline{DF}$; trans. $\overline{AB}$	1. Given
2. $\angle ACE \cong \angle ADF$	2. If 2 $\parallel$ lines are cut by a trans., then the corresponding $\angle$s are $\cong$.
3. $\overline{CX}$ bisects $\angle ACE$ $\overline{DE}$ bisects $\angle CDF$	3. Given
4. $\angle 1 \cong \angle 3$	4. If two $\angle$s are $\cong$, then their bisectors separate these $\angle$s into four $\cong \angle$s.

25. $x^2 - 9 = x(x - 1)$
$x^2 - 9 = x^2 - x$
$x = 9$

29. If two lines are cut by a transversal so that the alternate exterior angles are congruent, then these lines are parallel.
Given: Lines l and m and trans t;
$\angle 1 \cong \angle 2$
Prove: $l \parallel m$

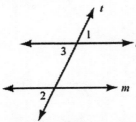

STATEMENTS	REASONS
1. Lines l and m and trans. t $\angle 1 \cong \angle 2$	1. Given
2. $\angle 1 \cong \angle 3$	2. If 2 lines intersect, the vertical $\angle$s formed are $\cong$.
3. $\angle 2 \cong \angle 3$	3. Transitive for $\cong$.
4. $l \parallel m$	4. If lines are cut by a trans. so that the corresponding angles are $\cong$, then these lines are $\parallel$.

33. Given: Line l and P not on l
Construct: The line through $P \parallel l$

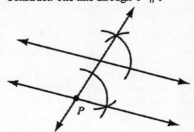

SECTION 2.4: The Angles of a Triangle

1. $m\angle C = 75°$

5. a. Underdetermined
b. Determined
c. Overdetermined

9. a. Equiangular Δ
b. Right Δ

13. $m\angle 1 = 122°$; $m\angle 2 = 58°$; $m\angle 5 = 72°$

17. $35°$

21. $360°$

25. $\qquad x + 4y = 180$
$\quad 2y + 2x - y - 40 = 180$

$x + 4y = 180 \qquad$ (Multiply by -2)
$2x + y = 220$

$\begin{aligned} -2x - 8y &= -360 \\ 2x + y &= 220 \\ \hline -7y &= -140 \\ y &= 20 \end{aligned}$
$x + 4(20) = 180$
$x + 100 = 180$
$\qquad x = 100$
$m\angle 2 = 80°$; $m\angle 3 = 40°$ $\therefore m\angle 5 = 60°$

29. a.

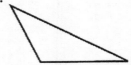

b. It is not possible to draw an equilateral right triangle.

33. $x + 2x + 33 = 180$
$\qquad 3x = 147$
$\qquad\quad x = 49$

$m\angle N = 49°$; $m\angle P = 98°$

37. 75°

41. The measure of an exterior angle of a triangle equals the sum of measures of the two nonadjacent interior angles.

Given: $\triangle ABC$ with ext. $\angle BCD$
Prove: $m\angle BCD = m\angle A + m\angle B$

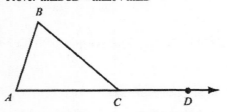

STATEMENTS	REASONS
1. $\triangle ABC$ with ext. $\angle BCD$	1. Given
2. $m\angle A + m\angle B + m\angle BCD = 180$	2. The sum of the measures of the $\angle$s in a $\triangle$ is 180.
3. $\angle BCA$ is supp. to $\angle BCD$	3. If the exterior sides of two adjacent $\angle$s form a straight line, then the angles are supp.
4. $m\angle BCA + m\angle BCD = 180$	4. If two $\angle$s are supp., then the sum of their measures is 180.
5. $m\angle BCA + m\angle BCD = m\angle A + m\angle B + m\angle BCA$	5. Substitution
6. $m\angle BCD = m\angle A + m\angle B$	6. Subtraction Prop. of Equality

45. $2b = m\angle M + 2a$
$\qquad (m\angle RPM = m\angle M + m\angle MNP)$
$\therefore m\angle M = 2b - 2a$
$b = 42 + a$
$\qquad (m\angle QPR = m\angle Q + m\angle QNP)$
$m\angle M = 2(42 + a) - 2a \qquad$ (Substitution)
$m\angle M = 84 - 2a + 2a$
$m\angle M = 84°$

SECTION 2.5: Convex Polygons

1. Increase

For #5 use $D = \dfrac{n(n-3)}{2}$

5. a. 5

    b. 35

For #9 use $I = \dfrac{180(n-2)}{n}$

9. a. 90°

    b. 150°

For #13 use $S = 180(n-2)$

13. a. 7

    b. 9

For #17 use $n = \dfrac{360}{E}$

17. a. 15

    b. 20

21.

25.

29. Given: Quad. *RSTV* with diagonals $\overline{RT}$ and $\overline{SV}$ intersecting at *W*

Prove: $m\angle 1 + m\angle 2 = m\angle 3 + m\angle 4$

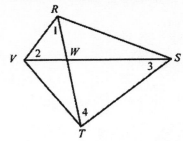

STATEMENTS	REASONS
1. Quad *RSTV* with diagonals $\overline{RT}$ and $\overline{SV}$ intersecting at *W*.	**1.** Given
2. $m\angle RWS = m\angle 1 + m\angle 2$	**2.** The measure of an ext. $\angle$ of a $\triangle$ equals the sum of the measures of the non-adjacent interior $\angle$s of the $\triangle$.
3. $m\angle RWS = m\angle 3 + m\angle 4$	**3.** Same as (2)
4. $m\angle 1 + m\angle 2 = m\angle 3 + m\angle 4$	**4.** Substitution

33. $36°$

For #37 use $I = \dfrac{180(n-2)}{n}$

37. $150°$

41. $221°$

SECTION 2.6: Symmetry and Transformations

1. M, T, X

5. (a), (c)

9. MOM

13. a.

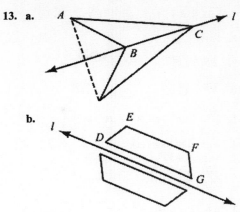

    b.

17. WHIM

21. WOW

25. 62,365 kilowatt hours

29. (b), (c)

CHAPTER REVIEW

1. a. $\overline{BC} \parallel \overline{AD}$

   b. $\overline{AB} \parallel \overline{CD}$

2. $m\angle 3 = 110°$

3. $2x + 17 = 5x - 94$
$111 = 3x$
$37 = x$

4. $m\angle A = 50°$ (corresponds to $\angle DCE$)
$\therefore\ m\angle BCA = 55$
$m\angle BCD = 75$ and $m\angle D = 75$
$m\angle DEF = 50 + 75 = 125$

5. $130 + 2x + y = 180$
$150 + 2x - y = 180$

$2x + y = 50$
$\underline{2x - y = 30}$
$4x = 80$
$x = 20$

$130 + 2(20) + y = 180$
$130 + 40 + y = 180$
$y = 10$

6. $2x + 15 = x + 45$
$x = 30$

$3y + 30 + 45 = 180$
$3y + 75 = 180$
$3y = 105$
$y = 35$

7. $\overline{AE} \parallel \overline{BF}$

8. None

9. $\overline{BE} \parallel \overline{CF}$

10. $\overline{BE} \parallel \overline{CF}$

11. $\overline{AC} \parallel \overline{DF}$ and $\overline{AE} \parallel \overline{BF}$

12. $x = 120$ (corr. $\angle$);
$x = y + 50$
$120 = y + 50$
$y = 70$

13. $x = 32; \; y = 30$

14. $2x - y = 3x + 2y$
$-1x - 3y = 2$ (Multiply by 3)
$3x - y = 80$

$-3x - 9 = 0$
$\underline{3x - y = 80}$
$-10y = 80$
$y = -8$

$3x + 8 = 80$
$3x = 72$
$x = 24$

15. $2a + 2b + 100 = 180$
$2a + 2b = 80$
$a + b = 40$ ($\div$ by 2)
$(a + b) + x = 180$
$40 + x = 180$
$x = 140$

16. $x^2 - 12 = x(x - 2)$
$x^2 - 12 = x^2 - 2x$
$-12 = -2x$
$x = 6$

17. $x^2 - 3x + 4 + 17x - x^2 - 5 = 111$
$14x - 1 = 222$
$14x = 112$
$x = 8$
$m\angle 3 = 69°$; $m\angle 4 = 67°$; $m\angle 5 = 44°$

18. $3x + y + 5x + 10 = 180$
$3x + y = 5y + 20$
$3x + y = 170$ (Multiply by 4)
$3x - 4y = 20$

$32x + 4y = 680$
$\underline{3x - y = 80}$
$35x = 700$
$x = 20$

$8(20) + y = 170$
$160 + y = 170$
$y = 10$

$m\angle C = 5(10) + 20 = 70°$ $\therefore m\angle B = 110°$

19. S

20. N

21. N

22. S

23. S

24. A

25.

Number of sides	8	12	20	15	10	16	180
Measure of each ext. $\angle$	45	30	18	24	36	22.5	2
Measure of each int. $\angle$	135	150	162	156	144	157.5	178
Number of diagonals	20	54	170	90	35	104	15,930

26.

27.

28. Not possible

29.

30. Statement: If 2 angles are right angles, then the angles are congruent.
Converse: If 2 angles are congruent, then the angles are right angles.
Inverse: If 2 angles are not right angles, then angles are not congruent.
Contrapositive: If 2 angles are not congruent, then the angles are not right angles.

31. Statement: If it is not raining, then I am happy.
Converse: If I am happy, then it is not raining.
Inverse: If it is raining, then I am not happy.
Contrapositive: If I am not happy, then it is raining.

32. Contrapositive

33. Given: $\overline{AB} \parallel \overline{CF}$
$\angle 2 \cong \angle 3$
Prove: $\angle 1 \cong \angle 3$

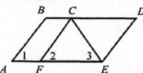

STATEMENTS	REASONS
1. $\overline{AB} \parallel \overline{CF}$	**1.** Given
2. $\angle 1 \cong \angle 2$	**2.** If 2 ∥ lines are cut by a trans., then the corresponding ∠s are ≅.
3. $\angle 2 \cong \angle 3$	**3.** Given
4. $\angle 1 \cong \angle 3$	**4.** Transitive Prop. of Congruence

34. Given: $\angle 1$ is comp. to $\angle 2$
$\angle 2$ is comp. to $\angle 3$
Prove: $\overline{BD} \parallel \overline{AE}$

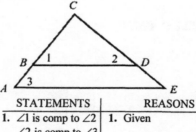

STATEMENTS	REASONS
1. $\angle 1$ is comp to $\angle 2$ $\angle 2$ is comp to $\angle 3$	**1.** Given
2. $\angle 1 \cong \angle 3$	**2.** If 2 ∠s are comp. to the same ∠, then these ∠s are ≅.
3. $\overline{BD} \parallel \overline{AE}$	**3.** If 2 lines are cut by a trans. so that the corresponding ∠s are ≅, then the lines are ∥.

35. Given: $\overline{BE} \perp \overline{DA}$
$\overline{CD} \perp \overline{DA}$
Prove: $\angle 1 \cong \angle 2$

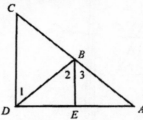

STATEMENTS	REASONS
1. $\overline{BE} \perp \overline{DA}$ $\overline{CD} \perp \overline{DA}$	**1.** Given
2. $\overline{BE} \parallel \overline{CD}$	**2.** If 2 lines are each ⊥ to a 3rd line, then these lines are parallel to each other.
3. $\angle 1 \cong \angle 2$	**3.** If 2 ∥ lines are cut by a trans., then the alternate interior ∠s are ≅.

36. Given: $\angle A \cong \angle C$
$\overline{DC} \parallel \overline{AB}$
Prove: $\overline{DA} \parallel \overline{CB}$

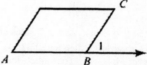

STATEMENTS	REASONS
1. $\angle A \cong \angle C$	**1.** Given
2. $\overline{DC} \parallel \overline{AB}$	**2.** Given
3. $\angle C \cong \angle 1$	**3.** If 2 ∥ lines are cut by a trans., then the alt. int. ∠s are congruent.
4. $\angle A \cong \angle 1$	**4.** Transitive Prop. of Congruence
5. $\overline{DA} \parallel \overline{CB}$	**5.** If 2 lines are cut by a trans. so that corr. ∠s are ≅, then these lines are ∥.

37. Assume $x = -3$.

38. Assume the sides opposite these angles are ≅.

39. Given: $m \not\parallel n$

Prove: $\angle 1 \not\cong \angle 2$

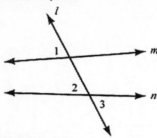

Indirect Proof:

Assume that $\angle 1 \cong \angle 2$. Then $m \parallel n$ since congruent corr. Angles are formed. But this contradicts our hypothesis. Therefore, our assumption must be false and $\angle 1 \not\cong \angle 2$.

40. Given: $\angle 1 \not\cong \angle 3$

Prove: $m \not\parallel n$

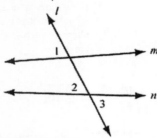

Indirect Proof:

Assume that $m \parallel n$. Then $\angle 1 \cong \angle 3$ since alt. ext. angles are congruent when parallel lines are cut by a transversal. But this contradicts the given fact that $\angle 1 \not\cong \angle 3$. Therefore, our assumption must be false and it follows that $m \not\parallel n$.

41. Given: $\triangle ABC$

Construct: The line through C parallel to $\overline{AB}$.

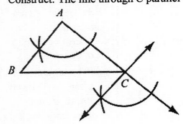

42. Given: $\overline{AB}$

Construct: An equilateral triangle ABC with side $\overline{AB}$.

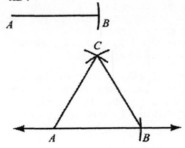

43. **a.** B, H, W

b. H, S

44. **a.** Isosceles triangle, Circle, Regular pentagon

b. Circle

45. Congruent

46. **a.**

b.

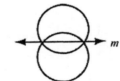

47. 90°

CHAPTER TEST

1. **a.** $\angle 5$

   b. $\angle 3$

2. **a.** $m\angle 2 + m\angle 8 = 68 + 112 = 180°$
   $m\angle 6 + m\angle 9 = 68 + 110 = 178°$
   $r \parallel s$

   b. $m\angle 2 + m\angle 8 = 68 + 112 = 180°$
   $m\angle 6 + m\angle 9 = 68 + 110 = 178°$
   $l \not\parallel m$

3. not Q

4. $\angle R$ and $\angle S$ are not both right angles.

5. a. If $r \parallel s$ and $s \parallel t$,

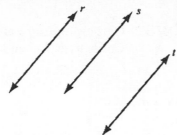

then $r \parallel t$.

b. If $a \perp b$ and $b \perp c$,

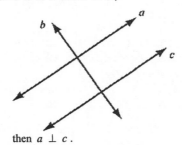

then $a \perp c$.

6.

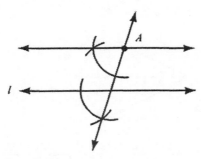

7. a. $65 + 79 + x = 180$
$144 + x = 180$
$x = 36$

$m\angle B = 36°$

b. $2x + x + 2x + 15 = 180$
$5x + 15 = 180$
$5x = 165$
$x = 33$

$m\angle B = 33°$

8. a. Pentagon

b. Use $D = \dfrac{n(n-3)}{2}$; 5

9. a. Equiangular hexagon

b. Use $I = \dfrac{180(n-2)}{n}$; 120°

10. A: line; D: line; N: point; O: both; X: both

11. a. Reflection

b. Slide

c. Rotation

12. $\angle 1 \cong \angle 2$ and $\angle 4 \cong \angle 3$
$m\angle C = 61°$

13. $x + 28 = 2x - 26$
$x = 54$

14. $m\angle 3 = 80°$ so $m\angle 4 = 50°$

15. $m\angle 3 = 63°$
$m\angle 1 = x$
$x + 2x + 63 = 180$
$3x = 117$
$x = 39$

$m\angle 1 = 39°$

16. 1. Given

2. $\angle 2 \cong \angle 3$

3. Transitive Prop. of Congruence

4. $l \parallel n$

17. Given: $\triangle MNQ$ with $m\angle N = 120°$
Prove: $\angle M$ and $\angle Q$ are not complementary

Indirect Proof:
Assume that $\angle M$ and $\angle Q$ are complementary.
By definition $m\angle M + m\angle Q = 90°$. Also,
$m\angle M + m\angle Q + m\angle N = 180°$ because these are
the three angles of $\triangle MNQ$. By substitution,
$90° + m\angle N = 180°$, so it follows that
$m\angle N = 90°$. But this leads to a contradiction
because it is given that $m\angle N = 120°$. The
assumption must be false, and it follows that
$\angle M$ and $\angle Q$ are not complementary.

18. 1. Given

2. 180°

3. $m\angle 1 + m\angle 2 + 90° = 180°$

4. 90°

S5. $\angle 1$ and $\angle 2$ are complementary.

R5. Definition of complementary angles.

Chapter 3: Triangles

SECTION 3.1: Congruent Triangles

1. $\angle A$; $\overline{AB}$; No; No

5. SAS

9. SSS

13. ASA

17. SSS

21. $\overline{AD} \cong \overline{EC}$

25. a. Given

    b. $\overline{AC} \cong \overline{AC}$

    c. SSS

29. Given: $\overline{AB} \perp \overline{BC}$ and
$\overline{AB} \perp \overline{BD}$;
also $\overline{BC} \cong \overline{BD}$
Prove: $\triangle ABC \cong \triangle ABD$

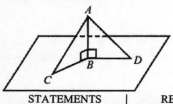

STATEMENTS	REASONS
1. $\overline{AB} \perp \overline{BC}$ and $\overline{AB} \perp \overline{BD}$	1. Given
2. $\angle ABC$ is a right $\angle$ and $\angle ABD$ is a right $\angle$	2. If 2 lines are $\perp$, then they meet to form a right $\angle$.
3. $\angle ABC \cong \angle ABD$	3. Any two right $\angle$s are $\cong$.
4. $\overline{BC} \cong \overline{BD}$	4. Given
5. $\overline{AB} \cong \overline{AB}$	5. Identity
6. $\triangle ABC \cong \triangle ABD$	6. SAS

33. Yes; SAS or SSS

37. a. $\triangle CBE$, $\triangle ADE$, $\triangle CDE$

    b. $\triangle ADC$

    c. $\triangle CBD$

SECTION 3.2: Corresponding Parts of Congruent Triangles

1. Given: $\angle 1$ and $\angle 2$ are right $\angle$ s.
$\overline{CA} \cong \overline{DA}$
Prove: $\triangle ABC \cong \triangle ABD$

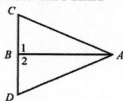

STATEMENTS	REASONS
1. $\angle 1$ and $\angle 2$ are right $\angle$s and $\overline{CA} \cong \overline{DA}$	1. Given
2. $\overline{AB} \cong \overline{AB}$	2. Identity
3. $\triangle ABC \cong \triangle ABD$	3. HL

5. Given: $\angle R$ and $\angle V$ are right $\angle$ s.
$\angle 1 \cong \angle 2$
Prove: $\triangle RST \cong \triangle VST$

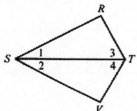

STATEMENTS	REASONS
1. $\angle R$ and $\angle V$ are right $\angle$s and $\angle 1 \cong \angle 2$	1. Given
2. $\angle R \cong \angle V$	2. All right $\angle$s are $\cong$.
3. $\overline{ST} \cong \overline{ST}$	3. Identity
4. $\triangle RST \cong \triangle VST$	4. AAS

9. $m\angle 2 = 48°$; $m\angle 3 = 48°$
$m\angle 5 = 42°$; $m\angle 6 = 42°$

13. Given: ∠s P and R are rt. ∠s
    M is the midpoint of $\overline{PR}$
    Prove: ∠N ≅ ∠Q

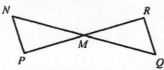

STATEMENTS	REASONS
1. ∠s P and R are rt. ∠s	1. Given
2. ∠P ≅ ∠R	2. All rt. ∠s are ≅.
3. M is the midpoint of $\overline{PR}$	3. Given
4. $\overline{PM} \cong \overline{MR}$	4. Midpoint of a segment forms 2 ≅ segments.
5. ∠NMP ≅ ∠QMR	5. If 2 lines intersect, the vertical angles formed are ≅.
6. △NPM ≅ △QRM	6. ASA
7. ∠N ≅ ∠Q	7. CPCTC

17. c = 5

21. c = √41

25. Given: E is the midpoint of $\overline{FG}$;
    $\overline{DF} \cong \overline{DG}$
    Prove: $\overline{DE} \perp \overline{FG}$

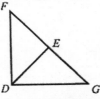

STATEMENTS	REASONS
1. E is the midpoint of $\overline{FG}$	1. Given
2. $\overline{FE} \cong \overline{EG}$	2. The midpoint of a segment forms 2 ≅ segments.
3. $\overline{DF} \cong \overline{DG}$	3. Given
4. $\overline{DE} \cong \overline{DE}$	4. Reflexive
5. △FDE ≅ △GDE	5. SSS
6. ∠DEF ≅ ∠DEG	6. CPCTC
7. $\overline{DE} \perp \overline{FG}$	7. If 2 lines meet to form ≅ adj. ∠s, then the lines are ⊥.

29. Given: $\overline{RW}$ bisects ∠SRU ;
    also $\overline{RS} \cong \overline{RU}$
    Prove: △TRU ≅ △VRS

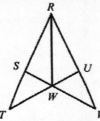

STATEMENTS	REASONS
1. $\overline{RW}$ bisects ∠RSU	1. Given
2. ∠SRW ≅ ∠URW	2. If a ray bisects an ∠ then 2 ≅ ∠s are formed.
3. $\overline{RS} \cong \overline{RU}$	3. Given
4. $\overline{RW} \cong \overline{RW}$	4. Identity
5. △RSW ≅ △RUW	5. SAS
6. ∠RSW ≅ ∠RUW	6. CPCTC
7. ∠TRV ≅ ∠VRS	7. Identity
8. △TRU ≅ △VRS	8. ASA

33. 751 feet

SECTION 3.3: Isosceles Triangles

1. Isosceles

5. m∠U = 69°

9. L = E (equivalent)

13. Underdetermined

17. Determined

21. m∠2 = 68° (Base ∠s in an isosceles △ are ≅)
    Also, m∠1 + m∠2 + m∠3 = 180
    m∠1 + 68 + 68 = 180
    m∠1 + 136 = 180
    m∠1 = 44°

25. Let the measure of the vertex angle be x. Then the measure of the base angles are each x + 12.
    $$x + (x+12) + (x+12) = 180$$
    $$3x + 24 = 180$$
    $$3x = 156$$
    $$x = 52$$
    m∠A = 52°; m∠B = 64°; m∠C = 64°

29. 12

33. 1. Given

   2. $\angle 3 \cong \angle 2$

   3. $\angle 1 \cong \angle 2$

   4. If 2 ∠s of one △ are ≅, then the opposite sides are ≅.

37. Given: Isosceles △*MNP* with vertex *P*;
            Isosceles △*MNQ* with vertex *Q*
      Prove: △*MQP* ≅ △*NQP*

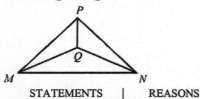

STATEMENTS	REASONS
1. Isosceles △*MNP* with vertex *P*	**1.** Given
2. $\overline{MP} \cong \overline{NP}$	**2.** An isosceles △ has 2 ≅ sides.
3. Isosceles △*MNQ* with vertex *Q*	**3.** Given
4. $\overline{MQ} \cong \overline{NQ}$	**4.** Same as (2)
5. $\overline{PQ} \cong \overline{PQ}$	**5.** Identity
6. △*MQP* ≅ △*NQP*	**6.** SSS

41. 75°

SECTION 3.4: Basic Constructions Justified

1.

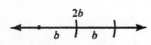

5.

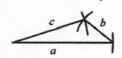

9.

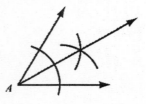

13.

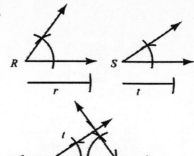

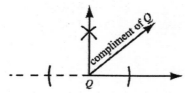

17.

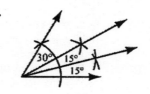

21.

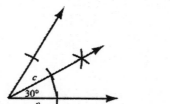

25.

29. Given: Line m, with point P on m
$\overline{PQ} \cong \overline{PR}$ (by construction)
$\overline{QS} \cong \overline{RS}$ (by construction)
Prove: $\overline{SP} \perp m$

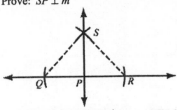

STATEMENTS	REASONS
1. Line m with point P on line m $\overline{PQ} \cong \overline{PR}$ and $\overline{QS} \cong \overline{RS}$	1. Given
2. $\overline{SP} \cong \overline{SP}$	2. Identity
3. $\triangle SPQ \cong \triangle SPR$	3. SSS
4. $\angle SPQ \cong \angle SPR$	4. CPCTC
5. $\overline{SP} \perp m$	5. If 2 lines intersect to form $\cong$ adjacent $\angle$s, then the lines are $\perp$.

33. 150°

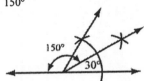

37. Yes

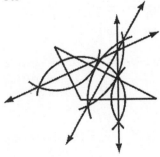

SECTION 3.5: Inequalities in a Triangle

1. False

5. True

9. True

13. **a.** Possible

   b. Not possible $(8 + 9 = 17)$

   c. Not possible $(8 + 9 < 18)$

17. Isosceles Obtuse triangle $(m\angle Z = 100°)$

21. Largest $\angle$ is 72° (two of these); smallest is 36°

25. **1.** $m\angle ABC > m\angle DBE$ and $m\angle CBD > m\angle EBF$

   3. Angle-Addition Postulate

   4. $m\angle ABD > m\angle DBF$

29. $BC < EF$

33. $x + 2 < y < 5x + 12$

37. The length of an altitude of a triangle that does not contain a right angle is less than the length of either side containing the same vertex as the altitude.
Given: $\triangle ABC$ with altitude $\overline{BD}$ to $\overline{AC}$
Prove: $BD < AB$ and $BD < BC$

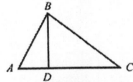

STATEMENTS	REASONS
1. $\triangle ABC$ with altitude $\overline{BD}$ to $\overline{AC}$	1. Given
2. $\overline{BD} \perp \overline{AC}$	2. Altitude of a $\triangle$ is the line segment drawn from a vertex $\perp$ to the opp. side.
3. $BD < AB$ and $BD < BC$	3. Shortest distance from a point to the line is the $\perp$ distance.

CHAPTER REVIEW

1. Given: $\angle AEB \cong \angle DEC$

$\overline{AE} \cong \overline{ED}$

Prove: $\triangle AEB \cong \triangle DEC$

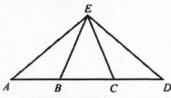

STATEMENTS	REASONS
1. $\angle AEB \cong \angle DEC$	1. Given
2. $\overline{AE} \cong \overline{ED}$	2. Given
3. $\angle A \cong \angle D$	3. If 2 sides of a △ are ≅, then the ∠s opposite these sides are also ≅.
4. $\triangle AEB \cong \triangle DEC$	4. ASA

2. Given: $\overline{AB} \cong \overline{EF}$

$\overline{AC} \cong \overline{DF}$

$\angle 1 \cong \angle 2$

Prove: $\angle B \cong \angle E$

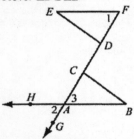

STATEMENTS	REASONS
1. $\overline{AB} \cong \overline{EF}$	1. Given
2. $\overline{AC} \cong \overline{DF}$; $\angle 1 \cong \angle 2$	2. Given
3. $\angle 2 \cong \angle 3$	3. If 2 lines intersect, then the vertical ∠s formed are ≅.
4. $\angle 1 \cong \angle 3$	4. Transitive Prop. for ≅.
5. $\triangle ABC \cong \triangle FED$	5. SAS
6. $\angle B \cong \angle E$	6. CPCTC

3. Given: $\overline{AD}$ bisects $\overline{BC}$

$\overline{AB} \perp \overline{BC}$

$\overline{DC} \perp \overline{BC}$

Prove: $\overline{AE} \cong \overline{ED}$

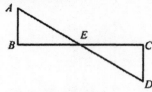

STATEMENTS	REASONS
1. $\overline{AB} \cong \overline{EF}$	1. Given
2. $\overline{AC} \cong \overline{DF}$; $\angle 1 \cong \angle 2$	2. Given
3. $\angle 2 \cong \angle 3$	3. If 2 lines intersect, then the vertical ∠s formed are ≅.
4. $\angle 1 \cong \angle 3$	4. Transitive Prop. for ≅.
5. $\triangle ABC \cong \triangle FED$	5. SAS
6. $\angle B \cong \angle E$	6. CPCTC

4. Given: $\overline{OA} \cong \overline{OB}$

$\overline{OC}$ is the median to $\overline{AB}$

Prove: $\overline{OC} \perp \overline{AB}$

STATEMENTS	REASONS
1. $\overline{OA} \cong \overline{OB}$	1. Given
2. $\overline{OC}$ is the median to $\overline{AB}$	2. Given
3. C is the midpoint of $\overline{AB}$	3. The median of a △ is a segment drawn from a vertex to the midpoint of the opp. side.
4. $\overline{AC} \cong \overline{CB}$	4. Midpoint of seg. form 2 ≅ segments.
5. $\overline{OC} \cong \overline{OC}$	5. Identity
6. $\triangle AOC \cong \triangle BOC$	6. SSS
7. $\angle OCA \cong \angle OCB$	7. CPCTC
8. $\overline{OC} \perp \overline{AB}$	8. If 2 lines meet to form ≅ adj. ∠s, then the lines are ⊥.

5. Given: $\overline{AB} \cong \overline{DE}$
   $\overline{AB} \parallel \overline{DE}$
   $\overline{AC} \cong \overline{DF}$

Prove: $\overline{BC} \parallel \overline{FE}$

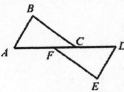

STATEMENTS	REASONS
1. $\overline{AB} \cong \overline{DE}$ and $\overline{AB} \parallel \overline{DE}$	1. Given
2. $\angle A \cong \angle D$	2. If 2 $\parallel$ lines are cut by a trans., then the alt. int. $\angle$s are $\cong$.
3. $\overline{AC} \cong \overline{DF}$	3. Given
4. $\triangle BAC \cong \triangle EDF$	4. SAS
5. $\angle BCA \cong \angle EFD$	5. CPCTC
6. $\overline{BC} \parallel \overline{FE}$	6. If 2 lines are cut by a trans. so that alt. int. $\angle$s are $\cong$, then the lines are $\parallel$.

6. Given: B is the midpoint of $\overline{AC}$
   $\overline{BD} \perp \overline{AC}$

Prove: $\triangle ADC$ is isosceles

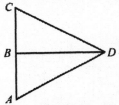

STATEMENTS	REASONS
1. B is the midpoint of $\overline{AC}$	1. Given
2. $\overline{CB} \cong \overline{BA}$	2. Midpoint of a segment forms 2 $\cong$ segments.
3. $\overline{BD} \perp \overline{AC}$	3. Given
4. $\angle DBC \cong \angle DBA$	4. If 2 lines are $\perp$, they meet to form $\cong$ adj. $\angle$s.
5. $\overline{BD} \cong \overline{BD}$	5. Identity
6. $\triangle CBD \cong \triangle ABD$	6. SAS
7. $\overline{DC} \cong \overline{DA}$	7. CPCTC
8. $\triangle ADC$ is isosceles	8. If a $\triangle$ has 2 $\cong$ sides, it is an isos. $\triangle$.

7. Given: $\overline{JM} \perp \overline{GM}$
   $\overline{GK} \perp \overline{KJ}$
   $\overline{GH} \cong \overline{HJ}$

Prove: $\overline{GM} \cong \overline{JK}$

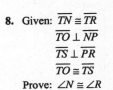

STATEMENTS	REASONS
1. $\overline{JM} \perp \overline{GM}$ and $\overline{GK} \perp \overline{KJ}$	1. Given
2. $\angle M$ is a rt. $\angle$ and $\angle K$ is a rt. $\angle$	2. If 2 lines are $\perp$, they meet to form a rt. $\angle$.
3. $\angle M \cong \angle K$	3. Any 2 rt. $\angle$s are $\cong$.
4. $\overline{GH} \cong \overline{HJ}$	4. Given
5. $\angle GHM \cong \angle JHK$	5. If 2 lines intersect, the vertical $\angle$s formed are $\cong$.
6. $\triangle GHM \cong \triangle JHK$	6. AAS
7. $\overline{GM} \cong \overline{JK}$	7. CPCTC

8. Given: $\overline{TN} \cong \overline{TR}$
   $\overline{TO} \perp \overline{NP}$
   $\overline{TS} \perp \overline{PR}$
   $\overline{TO} \cong \overline{TS}$

Prove: $\angle N \cong \angle R$

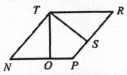

STATEMENTS	REASONS
1. $\overline{TN} \cong \overline{TR}$	1. Given
2. $\overline{TO} \perp \overline{NP}$; $\overline{TS} \perp \overline{PR}$	2. Given
3. $\angle TON$ is a rt. $\angle$ and $\angle TSR$ is a rt. $\angle$.	3. If 2 lines are $\perp$, they meet to form a rt. $\angle$.
4. $\overline{TO} \cong \overline{TS}$	4. Given
5. $\triangle TON \cong \triangle TSR$	5. HL
6. $\angle N \cong \angle R$	6. CPCTC

9. Given: $\overline{YZ}$ is the base of an isosceles triangle
   $\overline{XA} \parallel \overline{YZ}$

   Prove: $\angle 1 \cong \angle 2$

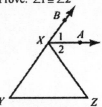

STATEMENTS	REASONS
1. $\overline{YZ}$ is the base of an isosceles △	1. Given
2. $\angle Y \cong \angle Z$	2. Base ∠s of an isos. △ are ≅.
3. $\overline{XA} \parallel \overline{YZ}$	3. Given
4. $\angle 1 \cong \angle Y$	4. If 2 ∥ lines are cut by a trans., then the corresp. ∠s are ≅.
5. $\angle 2 \cong \angle Z$	5. If 2 ∥ lines are cut by a trans., then the corresp. ∠s are ≅.
6. $\angle 1 \cong \angle 2$	6. Transitive Prop. for ≅.

10. Given: $\overline{AB} \parallel \overline{DC}$
    $\overline{AB} \cong \overline{DC}$
    C is the midpoint of $\overline{BE}$
    Prove: $\overline{AC} \parallel \overline{DE}$

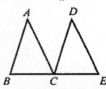

STATEMENTS	REASONS
1. $\overline{AB} \cong \overline{DE}$ and $\overline{AB} \parallel \overline{DC}$	1. Given
2. $\angle B \cong \angle DCE$	2. If 2 ∥ lines are cut by a trans., then the corresp. ∠s are ≅.
3. $\overline{AB} \cong \overline{DC}$	3. Given
4. C is the midpoint of $\overline{BE}$	4. Given
5. $\overline{BC} \cong \overline{CE}$	5. Midpoint of a seg. forms 2 ≅ segments.
6. △ABC ≅ △DCE	6. SAS
7. $\angle ACB \cong \angle E$	7. CPCTC
8. $\overline{AC} \parallel \overline{DE}$	8. If 2 lines are cut by a trans. so that the corresp. ∠s are ≅, then the lines are ∥.

11. Given: $\angle BAD \cong \angle CDA$
    $\overline{AB} \cong \overline{CD}$

    Prove: $\overline{AE} \cong \overline{ED}$

STATEMENTS	REASONS
1. $\angle BAD \cong \angle CDA$	1. Given
2. $\overline{AB} \cong \overline{CD}$	2. Given
3. $\overline{AD} \cong \overline{AD}$	3. Identity
4. △BAD ≅ △CDA	4. SAS
5. $\angle 1 \cong \angle 2$	5. CPCTC
6. $\overline{AE} \cong \overline{ED}$	6. If 2 ∠s of a triangle are ≅, then the sides opp. these ∠s are also ≅.

12. Given: $\overline{BE}$ is altitude to $\overline{AC}$
    $\overline{AD}$ is altitude to $\overline{CE}$
    $\overline{BC} \cong \overline{CD}$
    Prove: $\overline{BE} \cong \overline{AD}$

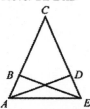

STATEMENTS	REASONS
1. $\overline{BE}$ is altitude to $\overline{AC}$ $\overline{AD}$ is altitude to $\overline{CE}$	1. Given
2. $\overline{BE} \perp \overline{AC}$ and $\overline{AD} \perp \overline{CE}$	2. An altitude is a line segment drawn from a vertex ⊥ to opp. side.
3. $\angle CBE$ is a rt. ∠ and $\angle CDA$ is a rt. ∠	3. If 2 lines are ⊥, they meet to form a rt. ∠.
4. $\angle CBE \cong \angle CDA$	4. Any 2 rt. ∠s are ≅.
5. $\overline{BC} \cong \overline{CD}$	5. Given
6. $\angle C \cong \angle C$	6. Identity
7. △CBE ≅ △CDA	7. ASA
8. $\overline{BE} \cong \overline{AD}$	8. CPCTC

13. Given: $\overline{AB} \cong \overline{CD}$
    $\angle BAD \cong \angle CDA$
    Prove: $\triangle AED$ is isosceles

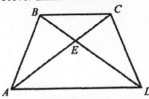

STATEMENTS	REASONS
1. $\overline{AB} \cong \overline{CD}$	1. Given
2. $\angle BAD \cong \angle CDA$	2. Given
3. $\overline{AD} \cong \overline{AD}$	3. Identity
4. $\triangle BAD \cong \triangle CDA$	4. SAS
5. $\angle CAD \cong \angle BDA$	5. CPCTC
6. $\overline{AE} \cong \overline{ED}$	6. If 2 $\angle$s of a $\triangle$ are $\cong$, then the sides opp. these $\angle$s are also $\cong$.
7. $\triangle AED$ is isosceles	7. If a $\triangle$ has 2 $\cong$ sides, it is an isosceles $\triangle$.

14. Given: $\overline{AC}$ bisects $\angle BAD$
    Prove: $AD > CD$

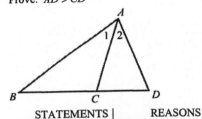

STATEMENTS	REASONS
1. $\overline{AC}$ bisects $\angle BAD$	1. Given
2. $m\angle 1 \cong m\angle 2$	2. If a ray bisects an $\angle$, it forms 2 $\angle$s of = measure.
3. $m\angle ACD > m\angle 1$	3. The measure of an ext. $\angle$ of a $\triangle$ is greater than the measure of either of the nonadjacent interior angles.
4. $m\angle ACD > m\angle 2$	4. Substitution
5. $AD > CD$	5. If the measure of one angle of a $\triangle$ is greater than the measure of a second angle, then the side which is opposite the 1st angle is longer than the side which is opp. the second angle.

15. a. $\overline{PR}$

    b. $\overline{PQ}$

16. $\overline{BC}$, $\overline{AC}$, $\overline{AB}$

17. $\angle R$, $\angle Q$, $\angle P$

18. $\overline{AD}$

19. (b)

20. 5 and 35

21. 20°

22. 115°

23. $3x + 10 = \frac{5}{2}x + 18$
    $\frac{1}{2}x = 8$
    $x = 16$
    $m\angle 4 = \frac{5}{2}(16) + 18 = 58°$
    $m\angle C = 64°$

24. $10 + x + 6 + 2x - 3 = 40$
    $3x + 13 = 40$
    $3x = 27$
    $x = 9$
    $AB = 10$; $BC = 15$; $AC = 15$: the triangle is isosceles.

25. Either $AB = BC$ or $AB = AC$ or $BC = AC$.
    If $AB = BC$, $y + 7 = 3y + 5$
    $ -2y = -2$
    $ y = 1$
    If $AB = AC$, $y + 7 = 9 - y$
    $ 2y = 2$
    $ y = 1$
    If $BC = AC$, $3y + 5 = 9 - y$
    $ 4y = 4$
    $ y = 1$
    If $y = 1$, $AB = 8$; $BC = 8$; $AC = 8$; the triangle is also equilateral.

26. If $m\angle 1 = 5x$, then the $m\angle 2 = 180 - 5x$
    $m\angle 4 = m\angle 2 = 180 - 5x$.
    But $m\angle 3 = m\angle 4$, therefore
    $2x + 12 = 180 - 5x$
    $7x = 168$
    $x = 24$
    $m\angle 2 = 180 - 5(24) = 60°$

27. Construct an angle that measures 75°.

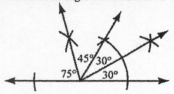

28. Construct a right triangle that has acute angle *A* and hypotenuse of length *c*.

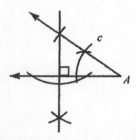

29. Construct another isosceles triangle in which the base angles are half as large as the given base angles.

CHAPTER TEST

1. **a.** Since $\triangle ABC \cong \triangle DEF$, then $m\angle A \cong m\angle D$, $m\angle B \cong m\angle E$ and $m\angle C \cong m\angle F$.

    $m\angle A \cong 37°$ and $m\angle E \cong 68°$

    $m\angle F = 180 - 37 - 68 = 75°$

    b. Since $\triangle ABC \cong \triangle DEF$, then $\overline{AB} \cong \overline{DE}$, $\overline{BC} \cong \overline{EF}$, and $\overline{AC} \cong \overline{DF}$.

    $AB = 7.3$, $BC = 4.7$, and $AC = 6.3$

    $EF = 4.7$ cm

2. **a.** $\overline{XY}$

    b. $\angle Y$

3. **a.** SAS

    b. ASA

4. Corresponding parts of congruent triangles are congruent.

5. **a.** No

    b. Yes

6. Yes

7. **a.** $c = 10$

    b. $b = \sqrt{c^2 - a^2}$
    $b = \sqrt{8^2 - 6^2}$
    $b = \sqrt{64 - 36}$
    $b = \sqrt{28} = 2\sqrt{7}$

8. **a.** $\overline{AM} \cong \overline{BM}$

    b. No

9. **a.** $m\angle V = 180 - 71 - 71 = 38°$

    b. $7x + 2 = 9(x - 2)$
    $7x + 2 = 9x - 18$
    $20 = 2x$
    $x = 10$
    $m\angle T = 7(10) + 2 = 72°$
    $m\angle U = 9(10 - 2) = 72°$
    $m\angle V = 180 - 72 - 72 = 36°$

10. **a.** $VU = 7.6$ in.

    b. $4x + 1 = 6x - 10$
    $11 = 2x$
    $x = \dfrac{11}{2}$
    $VT = 4\left(\dfrac{11}{2}\right) + 1 = 23$
    $TU = 2\left(\dfrac{11}{2}\right) = 11$
    $VU = 6\left(\dfrac{11}{2}\right) - 10 = 23$
    $P = 23 + 11 + 23 = 57$

11. **a.** Construct an angle that measures 60°.

    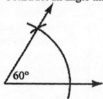

    b. Construct an angle that measures 60°.

    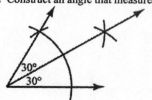

12.

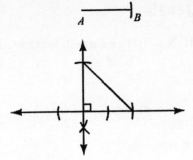

13. a. $\overline{BC}$

    b. $\overline{CA}$

14. $m\angle V > m\angle U > m\angle T$

15. $EB = \sqrt{(AE)^2 + (AB)^2}$

$EB = \sqrt{(4+3)^2 + 5^2}$

$EB = \sqrt{49 + 25} = \sqrt{74}$

$DC = \sqrt{(AD)^2 + (AC)^2}$

$DC = \sqrt{4^2 + (5+2)^2}$

$DC = \sqrt{16 + 49} = \sqrt{65}$

$EG > DC$ since $EB = \sqrt{74}$ and $DC = \sqrt{65}$.

16.

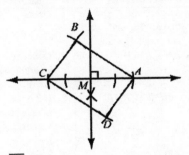

$\overline{DA}$

17.

STATEMENTS	REASONS
1. $\angle R$ and $\angle V$ are rt. $\angle$s	1. Given
2. $\angle R \cong \angle V$	2. All rt. $\angle$s are $\cong$
3. $\angle 1 \cong \angle 2$	3. Given
4. $\overline{ST} \cong \overline{ST}$	4. Identity
5. $\triangle RST \cong \triangle VST$	5. AAS

18. R1. Given

    R2. If 2 $\angle$s of a $\triangle$ are $\cong$, the opposite sides are $\cong$.

    S3. $\angle 1 \cong \angle 3$

    R4. ASA

    S5. $\overline{US} \cong \overline{UT}$

    S6. $\triangle STU$ is an isosceles triangle

Chapter 4: Quadrilaterals

SECTION 4.1: Properties of a Parallelogram

1. **a.** $AB = DC$

   b. $AD = BC$

5. In parallelogram $ABCD$, $AB = DC$.
   Therefore,
   $$3x + 2 = 5x - 2$$
   $$4 = 2x$$
   $$x = 2$$
   $$AB = DC = 8$$
   $$BC = AD = 9$$

9. $\overline{AC}$

13. True

17. The resulting quadrilateral appears to be a parallelogram.

21. **1.** Given

   2. $\overline{RV} \perp \overline{VT}$ and $\overline{ST} \perp \overline{VT}$

   3. $\overline{RV} \parallel \overline{ST}$

   4. $RSTV$ is a parallelogram

25. The opposite angles of a parallelogram are congruent.
   Given: Parallelogram $ABCD$
   Prove: $\angle BAD \cong \angle BCD$ and
   $\angle ABC \cong \angle ADC$

   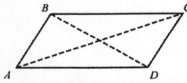

STATEMENTS	REASONS
1. Parallelogram $ABCD$	1. Given
2. Draw diagonal $\overline{BD}$	2. Through 2 points there is exactly one line.
3. $\triangle ABD \cong \triangle CDB$	3. A diagonal of a parallelogram separates it into 2 $\cong$ $\triangle$s.
4. $\angle BAD \cong \angle BCD$	4. CPCTC
5. Draw in diagonal $\overline{AC}$	5. Same as (2)
6. $\triangle ABC \cong \triangle CDA$	6. Same as (3)
7. $= m\angle A + m\angle B$	7. CPCTC

29. $\angle P$ is a right angle.

33. 255 mph

SECTION 4.2: The Parallelogram and Kite

1. **a.** Yes

   b. No

5. **a.** Kite

   b. Parallelogram

9. 6.18

13. 10

17. Parallel and congruent

21. Given: M-Q-T and P-Q-R so that
   $MNPQ$ and $QRST$ are parallelograms
   Prove: $\angle N \cong \angle S$

   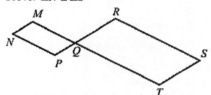

STATEMENTS	REASONS
1. M-Q-T and P-Q-R so that $MNPQ$ and $QRST$ are parallelograms.	1. Given
2. $\angle N \cong \angle MQP$	2. Opposite $\angle$s in a parallelogram are $\cong$.
3. $\angle MPQ \cong \angle RQT$	3. If 2 lines intersect, the vertical $\angle$s formed are $\cong$.
4. $\angle RQT \cong \angle S$	4. Same as (2)
5. $\angle N \cong \angle S$	5. Transitive Prop. for $\cong$.

25. If both pairs of opposite sides of a quadrilateral are congruent, then the quadrilateral is a parallelogram.

Given: Quad. $ABCD$ with
$\overline{AB} \cong \overline{CD}$ and $\overline{BC} \cong \overline{AD}$

Prove: $ABCD$ is a parallelogram

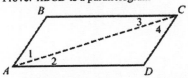

STATEMENTS	REASONS
1. Quad $ABCD$ with $\overline{AB} \cong \overline{CD}$ and $\overline{BC} \cong \overline{AD}$	1. Given
2. Draw in $\overline{AC}$	2. Through 2 points there is exactly one line.
3. $\overline{AC} \cong \overline{AC}$	3. Identity
4. $\triangle ABC \cong \triangle CDA$	4. SSS
5. $\angle 1 \cong \angle 4$ and $\angle 2 \cong \angle 3$	5. CPCTC
6. $\overline{AB} \parallel \overline{CD}$ and $\overline{BC} \parallel \overline{AD}$	6. If 2 lines are cut by a trans. so that alt. int. $\angle$s are $\cong$, then the lines are $\parallel$.
7. $ABCD$ is a parallelogram	7. If a quad. has both pairs of opposite sides $\parallel$, the quad. is a parallelogram.

29.
$$MN = \frac{1}{2} \cdot ST$$
$$2y - 3 = \frac{1}{2} \cdot 3y$$
$$2y - 3 = \frac{3}{2}y$$
$$\frac{1}{2}y = 3$$
$$y = 6$$
$$MN = 9$$
$$ST = 18$$

33.
$$m\angle B = 360 - m\angle A - m\angle C - m\angle D$$
$$= 360 - 30 - 30 - 30$$
$$= 270°$$

SECTION 4.3: The Rectangle, Square and Rhombus

1. $m\angle A = 60°$; $m\angle ABC = 120°$

5. The quadrilateral is a rhombus.

9. $2x + 7 = 3x + 2$
   $x = 5$
   $AD = BC = 3(5) + 4 = 19$

13. $QP = \sqrt{72} = 6\sqrt{2}$
    $MN = \sqrt{72} = 6\sqrt{2}$

17. $AD = \sqrt{34}$

21. True

25. (a)

29. A rectangle is a parallelogram. Therefore, the opposite angles are congruent and the consecutive angles are supplementary. If a rectangle has one right angle which measures 90°, the other three angles must also be 90°. Therefore, all the angles in a rectangle are right angles.

33. A diagonal of a rhombus bisects tow angles of the rhombus.

Given: $ABCD$ is a rhombus

Prove: $\overline{AC}$ bisects $\angle BAD$
and $\overline{CA}$ bisects $\angle BCD$

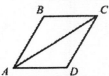

STATEMENTS	REASONS
1. $ABCD$ is a rhombus	1. Given
2. $\overline{AB} \cong \overline{BC} \cong \overline{CD} \cong \overline{AD}$	2. All sides of a rhombus are $\cong$.
3. $ABCD$ is a parallelogram	3. A rhombus is a parallelogram with 2 $\cong$ adj. sides.
4. $\angle B \cong \angle D$	4. Opposite angles of a parallelogram are $\cong$.
5. $\triangle ABC \cong \triangle ADC$	5. SAS
6. $\angle BAC \cong \angle DAC$ and $\angle BCA \cong \angle DCA$	6. CPCTC
7. $\overline{AC}$ bisects $\angle BAD$ and $\overline{CA}$ bisects $\angle BCD$	7. If a ray divides an $\angle$ into 2 $\cong$ $\angle$s, then the ray bisects the $\angle$.

37. If the midpoints of the sides of a rectangle are joined in order, then the quadrilateral formed is a rhombus.
    Given: Rect. *ABCD* with *M*, *N*, *O*, and *P*
    the midpoints of the sides.
    Prove: *MNOP* is a rhombus

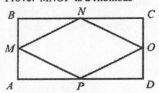

STATEMENTS	REASONS
1. Rect. *ABCD* with *M*, *N*, *O* and *P* the midpoints of the sides.	1. Given
2. *MNOP* is a parallelogram	2. From Exercise 36 of Section 4.2, when the midpoints of the consecutive sides of a quadrilateral are joined in order, the resulting quadrilateral is a parallelogram.
3. ∠s *A* and *B* are rt. ∠s	3. All angles of a rect. are rt. ∠s.
4. ∠*A* ≅ ∠*B*	4. Any 2 right ∠s are ≅.
5. $\overline{MB} \cong \overline{MA}$	5. The midpoint of a segment forms 2 ≅ segments.
6. $\overline{BC} \cong \overline{AD}$	6. Opposite sides of a parallelogram are ≅.
7. $\overline{BN} \cong \overline{AP}$	7. If two segments are ≅, then their midpoints separate these segments into four ≅ segments.
8. △*MBN* ≅ △*MAP*	8. SAS
9. $\overline{MN} \cong \overline{MP}$	9. CPCTC
10. *MNOP* is rhombus	10. If a parallelogram has 2 adj. sides ≅, then the parallelogram is a rhombus.

41. rhombus

SECTION 4.4: The Trapezoid

1. $m\angle D = 180 - 58 = 122°$
   $m\angle B = 180 - 125 = 55°$

5. The quadrilateral is a rhombus.

9. a. Yes

   b. No

13. $MN = \frac{1}{2}(AB + DC)$

    $9.5 = \frac{1}{2}(8.2 + DC)$

    $19 = 8.2 + DC$

    $DC = 10.8$

17. Given: *ABCD* is an isosceles trapezoid.
    Prove: △*ABE* is isosceles

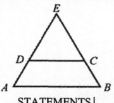

STATEMENTS	REASONS
1. *ABCD* is an isosceles trap.	1. Given
2. ∠*A* ≅ ∠*B*	2. Lower base angles of an isosceles trap. are ≅.
3. $\overline{EB} \cong \overline{EA}$	3. If 2 ∠s of a △ are ≅, then the sides opposite these ∠s are also ≅.
4. △*ABE* is isosceles	4. If a △ has 2 ≅ sides, it is an isosceles △.

21. $(QP)^2 = (MQ)^2 + (MP)^2$
$13^2 = 5^2 + (MP)^2$
$169 = 25 + (MP)^2$
$144 = (MP)^2$
$12 = MP$

25. $AB = BC$
$2x + 3 = x + 7$
$x = 4$
$EF = DE = 3x + 2$
$EF = 3(4) + 2 = 14$

29. If 2 consecutive angles of a quadrilateral are supplementary, the quadrilateral is a trapezoid.
Given: Quadrilateral *ABCD*
with $\angle A$ is supp. to $\angle B$
Prove: *ABCD* is a trapezoid

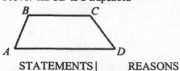

STATEMENTS	REASONS
1. Quad. *ABCD* with $\angle A$ is supp. to $\angle B$	1. Given
2. $\overline{BC} \parallel \overline{AD}$	2. If 2 lines are cut by a trans. so that the interior $\angle$s on the same side of the trans. are supp., then these lines are $\parallel$.
3. *ABCD* is a trapezoid	3. If a quad had 2 $\parallel$ sides, then the quad. is a trapezoid.

33. Given: $\overline{EF}$ is the median of trapezoid *ABCD*
Prove: $EF = \frac{1}{2}(AB + DC)$

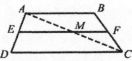

STATEMENTS	REASONS
1. $\overline{EF}$ is the median of trapezoid *ABCD*	1. Given
2. $\overline{AB} \parallel \overline{DC}$	2. Trapezoid has one pair of $\parallel$ sides.
3. $\overline{EF}$ is $\parallel$ to both $\overline{AB}$ and $\overline{DC}$	3. The median of a trap. is $\parallel$ to each base.
4. *E* is the midpoint of $\overline{AD}$ and *F* is the midpoint of $\overline{BC}$	4. The median of a trap. joins the midpoints of the nonparallel sides.
5. $\overline{AE} \cong \overline{ED}$	5. The midpoint of a segment forms 2 $\cong$ segments.
6. $\overline{AM} \cong \overline{MC}$	6. If 3 (or more) parallel lines intercept $\cong$ segments on one transversal, then they intercept $\cong$ segments on any transversal.
7. *M* is the midpoint of $\overline{AC}$	7. If a point divides a segment into 2 $\cong$ segments, then the point is the midpoint.
8. In $\triangle ADC$, $EM = \frac{1}{2}(DC)$ and in $\triangle ABC$, $MF = \frac{1}{2}(AB)$	8. The segment that joins the midpoints of two sides of a $\triangle$ is $\parallel$ to the third side and has a length equal to one-half the length of the third side.
9. $EM + MF = \frac{1}{2}(AB) + \frac{1}{2}(DC)$	9. Addition Property of Equality
10. $EM + MF = \frac{1}{2}(AB + DC)$	10. Distrubutive Property
11. $EF = EM + MF$	11. Segment Addition Postulate
12. $EF = \frac{1}{2}(AB + DC)$	12. Substitution

37. **a.** $AS = 3$ ft

     b. $VD = 12$ ft

     c. $CD = 13$ ft

     d. $DE = \sqrt{73}$ ft

CHAPTER REVIEW

1. A

2. S

3. N

4. S

5. S

6. A

7. A

8. A

9. A

10. N

11. S

12. N

13. $2(2x+3) + 2(5x-4) = 96$
$$4x + 6 + 10x - 8 = 96$$
$$14x - 2 = 06$$
$$14x = 98$$
$$x = 7$$
$AB = DC = 2(7) + 3 = 17$
$AD = BC = 5(7) - 4 = 31$

14. $2x + 6 + x + 24 = 180$
$$3x + 30 = 180$$
$$3x = 150$$
$$x = 50$$
$m\angle C = m\angle A = 2(50) + 6 = 106°$

15. The sides of a parallelogram measure 13 since $5^2 + 12^2 = (\text{side})^2$. Perimeter is 52.

16. $4x = 2x + 50$
$$2x = 50$$
$$x = 25$$
$m\angle M = 4(25) = 100°$
$m\angle P = 180 - 100 = 80°$

17. $\overline{PN}$

18. Kite

19. $m\angle G = m\angle F = 180 - 108 = 72°$
$m\angle E = 108°$

20. Median $= \dfrac{1}{2}(12.3 + 17.5)$
$$= \dfrac{1}{2}(29.8)$$
$$= 14.9$$

21. $15 = \dfrac{1}{2}(3x + 2 + 2x - 7)$
$$30 = 5x - 5$$
$$35 = 5x$$
$$x = 7$$
$MN = 3(7) + 2 = 23$
$PO = 2(7) - 7 = 7$

22. If $\overline{FJ} \cong \overline{FH}$ and M and N are their midpoints, then $FM = NH$ or
$$2y + 3 = 5y - 9$$
$$-3y = -12$$
$$y = 4$$
$FM = 2(4) + 3 = 11$
$FN = NH = 5(4) - 9 = 11$
$JH = 2(4) = 8$
The perimeter of $\triangle FMN = 26$.

23. Since M and N are midpoints, $\overline{MN} \parallel \overline{JH}$ and $MN = \dfrac{1}{2} \cdot JH$. There fore, $MN = 6$,
$$m\angle FMN = 80° \text{ and } m\angle FNM = 40°.$$

24. Since M and N are midpoints, $MN = \dfrac{1}{2} \cdot JH$.

Therefore $x^2 + 6 = \dfrac{1}{2} \cdot 2x(x + 2)$
$$x^2 + 6 = x(x + 2)$$
$$x^2 + 6 = x^2 + 2x$$
$$6 = 2x$$
$$x = 3$$
$MN = 15$
$JH = 30$

25. Given: *ABCD* is a parallelogram
  $\overline{AF} \cong \overline{CE}$
Prove: $\overline{DF} \parallel \overline{EB}$

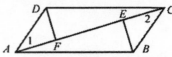

STATEMENTS	REASONS
1. *ABCD* is a parallelogram	1. Given
2. $\overline{AD} \cong \overline{CB}$	2. Opp. sides of a parallelogram are ≅.
3. $\overline{AD} \parallel \overline{CB}$	3. Opp. sides of a parallelogram are ∥.
4. ∠1 ≅ ∠2	4. If 2 ∥ lines are cut by a trans., then the alt. int. ∠s are ≅.
5. $\overline{AF} \cong \overline{CE}$	5. Given
6. △*DAF* ≅ △*BCE*	6. SAS
7. ∠*DFA* ≅ ∠*BEC*	7. CPCTC
8. $\overline{DF} \parallel \overline{EB}$	8. If 2 lines are cut by a trans. so that alt. ex. ∠s are ≅, then the lines are ∥.

26. Given: *ABEF* is a rect., *BCDE* is a rect.
  $\overline{FE} \cong \overline{ED}$
Prove: $\overline{AE} \cong \overline{BD}$ and $\overline{AE} \parallel \overline{BD}$

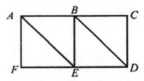

STATEMENTS	REASONS
1. *ABEF* is a rect.	1. Given
2. *ABEF* is a parallelogram	2. A rect. is a parallogram with a rt. ∠.
3. $\overline{AF} \cong \overline{BE}$	3. Opp. sides of a parallelogram are ≅.
4. *BCDE* is a rect.	4. Given
5. ∠*F* and ∠*BED* are rt. ∠s	5. Same as (2)
6. ∠*F* ≅ ∠*BED*	6. Any 2 rt. ∠s are ≅.
7. $\overline{FE} \cong \overline{ED}$	7. Given
8. △*AFE* ≅ △*BED*	8. SAS
9. $\overline{AE} \cong \overline{BD}$	9. CPCTC
10. ∠*AEF* ≅ ∠*BDE*	10. CPCTC
11. $\overline{AE} \parallel \overline{BD}$	11. If lines are cut by a trans. so that the corresp. ∠s are ≅, then the lines are ∥.

27. Given: $\overline{DE}$ is a median in △*ADC*
  $\overline{BE} \cong \overline{FD}$ and $\overline{EF} \cong \overline{FD}$
Prove: *ABCF* is a parallelogram

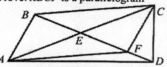

STATEMENTS	REASONS
1. $\overline{DE}$ is a median of △*ADC*	1. Given
2. *E* is the midpoint of $\overline{AC}$	2. Median of a △ is a line segment drawn from a vertex to the midpoint of the opp. side.
3. $\overline{AE} \cong \overline{EC}$	3. Midpoint of a segment forms 2 ≅ segments.
4. $\overline{BE} \cong \overline{FD}$ and $\overline{EF} \cong \overline{FD}$	4. Given
5. $\overline{BE} \cong \overline{EF}$	5. Transitive Prop. for ≅
6. *ABCF* is a parallelogram	6. If the diagonals of a quad. bisect each other then the quad is a parallelogram.

28. Given: △*FAB* ≅ △*HCD*
  △*EAD* ≅ △*GCB*
Prove: *ABCD* is a parallelogram

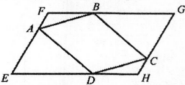

STATEMENTS	REASONS
1. △*FAB* ≅ △*HCD*	1. Given
2. $\overline{AB} \cong \overline{DC}$	2. CPCTC
3. △*EAD* ≅ △*GCB*	3. Given
4. $\overline{AD} \cong \overline{BC}$	4. CPCTC
5. *ABCD* is a parallelogram	5. If a quad. has both pairs of opp. sides ≅, then the quad is a parallelogram.

29. Given: *ABCD* is a parallelogram

$\overline{DC} \cong \overline{BN}$

$\angle 3 \cong \angle 4$

Prove: *ABCD* is a rhombus

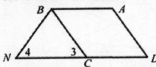

STATEMENTS	REASONS
1. *ABCD* is a parallelogram	**1.** Given
2. $\overline{DC} \cong \overline{BN}$	**2.** Given
3. $\angle 3 \cong \angle 4$	**3.** Given
4. $\overline{BN} \cong \overline{BC}$	**4.** If 2 $\angle$s of a $\triangle$ are $\cong$, then the sides opp. these $\angle$s are also $\cong$.
5. $\overline{DC} \cong \overline{BC}$	**5.** Transitive Prop. for $\cong$.
6. *ABCD* is a rhombus	**6.** If a parallelogram has 2 $\cong$ adj. sides, then the parallelogram is a rhombus

30. Given: $\triangle TWX$ is an isosceles with base $\overline{WX}$

$\overline{RY} \parallel \overline{WX}$

Prove: *RWXY* is an isosceles trapezoid

STATEMENTS	REASONS
1. $\triangle TWX$ is isosceles with base $\overline{WX}$	**1.** Given
2. $\angle W \cong \angle X$	**2.** Base $\angle$s of an isos. $\triangle$ are $\cong$.
3. $\overline{RY} \parallel \overline{WX}$	**3.** Given
4. $\angle TRY \cong \angle W$ and $\angle TYR \cong \angle X$	**4.** If 2 $\parallel$ lines are cut by a trans., then the corresp. $\angle$s are $\cong$.
5. $\angle TRY \cong \angle TYR$	**5.** Transitive Prop. for $\cong$.
6. $\overline{TR} \cong \overline{TY}$	**6.** If 2 $\angle$s of a $\triangle$ are $\cong$, then the side opp. these $\angle$s are also $\cong$.
7. $\overline{TW} \cong \overline{TX}$	**7.** Isosceles $\triangle$ has 2 $\cong$ sides.
8. $TR = TY$ and $TW = TX$	**8.** If 2 segments are $\cong$, then they are equal in length.
9. $TW = TR + RW$ and $TX = TY + YX$	**9.** Segment-Addition Post.
10. $TR + RW = TY + YX$	**10.** Substitution
11. $RW = YX$	**11.** Subtraction Prop. of Eq.
12. $\overline{RW} \cong \overline{YX}$	**12.** If segments are $\cong$ in length, then they are $\cong$.
13. *RWXY* is an isosceles trapezoid.	**13.** If a quad. has one pair of $\parallel$ sides and the non-parallel sides are $\cong$, then the quad. is an isos. trap.

31.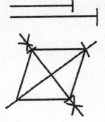

32. **a.** $\overline{AB} \perp \overline{BC}$

    b. $AC = 13$

33. **a.** $\overline{WY} \perp \overline{XZ}$

    b. $WY = 30$

34. **a.** Kites, rectangles, squares, rhombi, isosceles trapezoids

    b. Parallelograms, rectangles, squares, rhombi

35. **a.** Rhombus

    b. Kite

CHAPTER TEST

1. **a.** Congruent

    b. Supplementary

2. 18.8 cm

3. $CD = AB$. Let $x = AE$.
$$(AE)^2 + (DE)^2 = (AD)^2$$
$$x^2 + 4^2 = 5^2$$
$$x^2 = 9$$
$$x = 3$$
$$EB = AB - AE = 9 - 3 = 6$$

4. $m\angle S = 57°$
$m\angle R = 180 - 57 = 123°$
$\overline{VS}$ is longer.

5. $VT = 3x - 1$, $TS = 2x + 1$, $RS = 4(x - 2)$
$$3x - 1 = 4(x - 2)$$
$$x = 7$$

6. **a.** Kite

    b. Parallelogram

7. **a.** Altitude

    b. Rhombus

8. **a.** The line segments are parallel.

    b. $MN = \frac{1}{2}(BC)$

9. $MN = \frac{1}{2}(BC)$
$$7.6 = \frac{1}{2}(BC)$$
$$BC = 15.2 \text{ cm}$$

10. $MN = 3x - 11$
$BC = 4x + 24$
$$MN = \frac{1}{2}(BC)$$
$$3x - 11 = \frac{1}{2}(4x + 24)$$
$$x = 23$$

11. Let $x = AC$. $AD = 12$ and $DC = 5$
$$5^2 + 12^2 = x^2$$
$$x^2 = 169$$
$$x = 13$$
$$AC = 13$$
Or use the Pythagorean Triple (5, 12, 13).

12. **a.** $\overline{RV}$, $\overline{ST}$

    b. $\angle R$ and $\angle V$ (or $\angle S$ and $\angle T$)

13. $MN = \frac{1}{2}(RS + VT)$
$$= \frac{1}{2}(12.4 + 16.2)$$
$$= 14.3 \text{ in.}$$

14. $VT = 2x + 9$, $MN = 6x - 13$, $RS = 15$
$$MN = \frac{1}{2}(RS + VT)$$
$$6x - 13 = \frac{1}{2}(15 + 2x + 9)$$
$$6x - 13 = x + 12$$
$$5x = 25$$
$$x = 5$$

15. **S1.** Kite $ABCD$; $\overline{AB} \cong \overline{AD}$ and $\overline{BC} \cong \overline{DC}$

    R1. Given

    S3. $\overline{AC} \cong \overline{AC}$

    R4. SSS

    S5. $\angle B \cong \angle D$

    R5. CPCTC

16. **S1.** Trap. $ABCD$ with $\overline{AB} \parallel \overline{DC}$ and $\overline{AD} \cong \overline{BC}$

    R2. Congruent

    R3. Identity

    R4. SAS

    S5. $\overline{AC} \cong \overline{DB}$

Chapter 5: Similar Triangles

SECTION 5.1: Ratios, Rates and Proportions

1. **a.** $\dfrac{12}{15}=\dfrac{4}{5}$

    b. $\dfrac{12\text{ inches}}{15\text{ inches}}=\dfrac{4}{5}$

    c. $\dfrac{1\text{ foot}}{18\text{ inches}}=\dfrac{12\text{ inches}}{18\text{ inches}}=\dfrac{2}{3}$

    d. $\dfrac{1\text{ foot}}{18\text{ ounces}}$ is incommensurable

5. **a.** $12x=36$
    $\quad x=3$

    b. $21x=168$
    $\quad x=8$

9. **a.** $x^2=28$
    $\quad x=\pm\sqrt{28}=\pm2\sqrt{7}\approx\pm5.29$

    b. $x^2=18$
    $\quad x=\pm\sqrt{18}=\pm3\sqrt{2}\approx\pm4.24$

13. **a.** $\qquad 3(x+1)=2x^2$
    $\qquad 3x+3=2x^2$
    $2x^2-3x-3=0$
    $a=2;\ b=-3;\ c=-3$
    $x=\dfrac{-b\pm\sqrt{b^2-4ac}}{2a}$
    $x=\dfrac{-(-3)\pm\sqrt{(-3)^2-4(2)(-3)}}{2(2)}$
    $x=\dfrac{3\pm\sqrt{9+24}}{4}$
    $x=\dfrac{3\pm\sqrt{33}}{4}\approx2.19\text{ or }-0.69$

    b. $5(x+1)=2x(x-1)$
    $\quad 5x+1=2x^2-2x$
    $\qquad 0=2x^2-7x-5$
    $a=2;\ b=-7;\ c=-5$
    $x=\dfrac{-b\pm\sqrt{b^2-4ac}}{2a}$
    $x=\dfrac{-(-7)\pm\sqrt{(-7)^2-4(2)(-5)}}{2(2)}$
    $x=\dfrac{7\pm\sqrt{49+40}}{4}$
    $x=\dfrac{7\pm\sqrt{89}}{4}\approx4.19\text{ or }-0.61$

17. $\dfrac{4\text{ eggs}}{3\text{ cups of milk}}=\dfrac{14\text{ eggs}}{x\text{ cups of milk}}$
    $\qquad 4x=42$
    $\qquad x=\dfrac{42}{4}\text{ or }10\dfrac{1}{2}\text{ cups of milk}$

21. **a.** $\dfrac{BD}{AD}=\dfrac{AD}{DC}$
    $\quad \dfrac{6}{AD}=\dfrac{AD}{8}$
    $\quad (AD)^2=48$
    $\quad AD=\sqrt{48}=4\sqrt{3}\approx6.93$

    b. $\dfrac{BD}{AD}=\dfrac{AD}{DC}$
    $\quad \dfrac{BD}{6}=\dfrac{6}{8}$
    $\quad 8(BD)=36$
    $\quad BD=\dfrac{36}{8}=4\dfrac{1}{2}$

25. Let the first angle have measure x so that the complementary angle has measure $90-x$. Then
    $\dfrac{x}{90-x}=\dfrac{4}{5}$
    $5x=4(90-x)$
    $5x=360-4x$
    $9x=360$
    $x=40;\ 90-x=50$
    The angles measure $40°$ and $50°$.
    Alternate method: Let the measures of the two angles be $4x$ and $5x$. Then
    $4x+5x=90$
    $\quad 9x=90$
    $\quad x=10;\ 4x=40;\ 5x=50$
    The angles measure $40°$ and $50°$.

29. $\dfrac{7}{3}=\dfrac{6}{YZ}$
    $7\cdot YZ=18$
    $\quad YZ=2\dfrac{4}{7}\approx2.57$

33. $\dfrac{1\text{ in.}}{3\text{ ft}}=\dfrac{x\text{ in.}}{12\text{ ft}}$
    $\quad 3x=12$
    $\quad x=4\text{ in.}$
    $\dfrac{1\text{ in.}}{3\text{ ft}}=\dfrac{y\text{ in.}}{14\text{ ft}}$
    $\quad 3y=14$
    $\quad y=4\dfrac{2}{3}\text{ in.}$

    The blue print should be 4 in. by $4\dfrac{2}{3}$ in.

SECTION 5.2: Similar Polygons

1. **a.** Congruent

   b. Proportional

5. **a.** $\triangle ABC \sim \triangle XTN$

   b. $\triangle ACB \sim \triangle NXT$

9. **a.** $m\angle N = 82°$

   b. $m\angle N = 42°$

   c. $\dfrac{NP}{RS} = \dfrac{NM}{RQ}$

   $\dfrac{NP}{7} = \dfrac{9}{6}$ or $\dfrac{3}{2}$

   $2 \cdot NP = 21$

   $NP = 10\dfrac{1}{2}$

   d. $\dfrac{MP}{QS} = \dfrac{MN}{RQ}$

   $\dfrac{12}{QS} = \dfrac{9}{6}$ or $\dfrac{3}{2}$

   $3 \cdot QS = 24$

   $QS = 8$

13. $\dfrac{HK}{KF} = \dfrac{HJ}{FG}$

    $\dfrac{4}{FG} = \dfrac{6}{8}$ or $\dfrac{3}{4}$

    $3 \cdot FG = 16$

    $FG = 5\dfrac{1}{3}$

17. $\dfrac{AB}{HJ} = \dfrac{BC}{JK}$

    $\dfrac{n}{n+3} = \dfrac{5}{10}$ or $\dfrac{1}{2}$

    $n + 3 = 2n$

    $n = 3$

21. Let $BC = x$; then $CE = x$ and $CA = x + 6$.

    $\dfrac{4}{x} = \dfrac{6}{x+6}$

    $6x = 4(x + 6)$

    $6x = 4x + 24$

    $2x = 24$

    $x = 12;\ BC = 12$

25. Quadrilateral $MNPQ \sim$ quadrilateral $WXYZ$

    $\dfrac{6}{9} = \dfrac{50}{n}$

    $6n = 9 \cdot 50$

    $6n = 450$

    $n = 75$

29. Let $x =$ the boy's height.

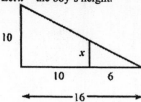

    $\dfrac{6}{x} = \dfrac{16}{10}$ or $\dfrac{8}{5}$

    $8x = 30$

    $x = \dfrac{30}{8}$ or $3\dfrac{3}{4}$, the boy is 3 ft 9 in.

33. No. The sides of quadrilateral $ABCD$ are not proportional to quadrilateral $DCFE$.

SECTION 5.3: Proving Triangles Similar

1. CASTC

5. $SSS \sim$

9. $SAS \sim$

13. 1. Given

    2. Definition of midpoint

    3. If a line segment joins the midpoints of two sides of a $\triangle$, its length is $\dfrac{1}{2}$ the length of the third side.

    4. Division Prop. of Eq.

    5. Substitution

    6. $SSS \sim$

17. 1. $\angle H \cong \angle F$

    2. If two $\angle$s are vertical $\angle$s, then they are $\cong$.

    S3. $\triangle HJK \sim \triangle FGK$

    R3. AA

21. S1. $\overline{RS} \parallel \overline{UV}$

    R1. Given

    2. 2 $\parallel$ lines are cut by a transversal, alternate interior $\angle$s are $\cong$.

    3. $\triangle RST \sim \triangle VUT$

    S4. $\dfrac{RT}{VT} = \dfrac{RS}{VU}$

    R4. CSSTP

25. Let $DB = x$; then $AB = DB + AD = x + 4$.

$$\frac{AC}{DE} = \frac{AB}{DB}$$

$$\frac{10}{8} = \frac{x+4}{x}$$

$$8(x+4) = 10x$$

$$8x + 32 = 10x$$

$$2x = 32$$

$$x = 16$$

$$DB = 16$$

29. Since $\triangle ABF \sim \triangle CBD$, then.

$$m\angle C + m\angle B + m\angle AFB = 180°$$

$$45° + x + 4x = 180°$$

$$5x = 135°$$

$$x = 27°$$

33. "The lengths of the corresponding altitudes of similar triangles have the same ratio as the lengths of any pair of corresponding sides."

Given: $\triangle DEF \sim \triangle MNP$

$\overline{DG}$ and $\overline{MQ}$ are altitudes

Prove: $\dfrac{DG}{MQ} = \dfrac{DE}{MN}$

STATEMENTS	REASONS
1. $\triangle DEF \sim \triangle MNP$ $\overline{DG}$ and $\overline{MQ}$ are altitudes.	1. Given
2. $\overline{DG} \perp \overline{EF}$ and $\overline{MQ} \perp \overline{NP}$	2. An altitude is a segment drawn from a vertex $\perp$ to the opposite side.
3. $\angle DGE$ and $\angle MQN$ are rt. $\angle$s	3. $\perp$ lines form a rt. $\angle$.
4. $\angle DGE \cong \angle MQN$	4. Right $\angle$s are $\cong$.
5. $\angle E \cong \angle N$	5. If two $\triangle$s are $\sim$ then the corresponding $\angle$s are $\cong$.
6. $\triangle DGE \sim \triangle MQN$	6. AA
7. $\dfrac{DG}{MQ} = \dfrac{DE}{MN}$	7. Corresp. sides of $\sim$ $\triangle$s are proportional.

37.

$$\frac{XY}{YZ} = \frac{TY}{YW} = \frac{XT}{WZ}$$

$$\frac{120}{40} = \frac{XT}{50}$$

$$3 = \frac{XT}{50}$$

$$XT = 150 \text{ feet}$$

SECTION 5.4: The Pythagorean Theorem

1. $\triangle RST \sim \triangle RVS \sim \triangle SVT$

5.

$$\frac{RV}{6} = \frac{6}{8}$$

$$8 \cdot RV = 36$$

$$RV = 4.5$$

9. a. $(DF)^2 = (DE)^2 + (EF)^2$

$$17^2 = 15^2 + (EF)^2$$

$$289 = 225 + (EF)^2$$

$$(EF)^2 = 64$$

$$EF = 8$$

Or, (8, 15, 17) is a Pythagorean Triple; therefore $EF = 8$.

b. $(DF)^2 = (DE)^2 + (EF)^2$

$$12^2 = \left(8\sqrt{2}\right)^2 + (EF)^2$$

$$144 = 128 + (EF)^2$$

$$(EF)^2 = 16$$

$$EF = 4$$

13. Let c be the longest side.

a. $5^2 = 3^2 + 4^2$
Right $\triangle$

b. $6^2 < 4^2 + 5^2$
Acute $\triangle$

c. $\left(\sqrt{7}\right)^2 = 2^2 + \left(\sqrt{3}\right)^2$
Right $\triangle$

d. No $\triangle$

17. Let $x =$ the length of rope needed.

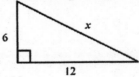

$$x^2 = 6^2 + 12^2$$

$$x^2 = 36 + 144$$

$$x^2 = 180$$

$$x = \sqrt{180} = 6\sqrt{5} \approx 13.4 \text{ meters}$$

21. Let $x =$ the length of the rectangle.

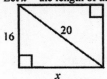

$20^2 = 16^2 + x^2$

$400 = 256 + x^2$

$x^2 = 144$

$x = 12$ cm

Or (12, 16, 20) is a multiple of the Pythagorean Triple (3, 4, 5).

25. Since the diagonals of a rhombus are perpendicular and since they bisect each other, one side of the right triangle has length 9 and the hypotenuse has length 12.

Let x represent the length of the other side of the right triangle.

$12^2 = x^2 + 9^2$

$144 = x^2 + 81$

$x^2 = 63$

$x = \sqrt{63} = 3\sqrt{7}$

The length of the other diagonal is $6\sqrt{7} \approx 15.87$ in.

29. In rt. $\triangle ACB$, $AC = 8$ using the triple (8, 15, 17). Since M and N are midpoints, $\overline{MN} \parallel \overline{AC}$. Since $\overline{MN} \parallel \overline{AC}$, $MN = \frac{1}{2} \cdot AC$. Therefore, $MN = 4$.

33. In rt. $\triangle RST$, if $RS = 6$ and $ST = 8$, then $RT = 10$ using the Pythagorean Triple (6, 8, 10). In rt. $\triangle RTU$, $RT = 10$ and $RU = 15$; let $UT = x$ and use the Pythagorean Theorem.

$15^2 = 10^2 + x^2$

$225 = 100 + x^2$

$x^2 = 125$

$x = \sqrt{125} = 5\sqrt{5} \approx 11.18$

37.

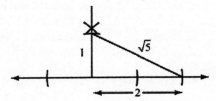

41. $TS = 13$ using the Pythagorean Triple (5, 12, 13).

$(RT)^2 = (RS)^2 + (TS)^2$

$(RT)^2 = 13^2 + 13^2$

$(RT)^2 = 169 + 169$

$(RT)^2 = 338$

$RT = \sqrt{338} = 13\sqrt{2} \approx 18.38$

SECTION 5.5: Special Right Triangles

1. a. $BC = a$

b. $AB = \sqrt{a^2 + a^2} = a\sqrt{2}$

5. $YZ = 8$ and $XY = 8\sqrt{2} \approx 11.31$

9. $DF = 5\sqrt{3} \approx 8.66$ and $FE = 10$

13. In right $\triangle HLK$, if $m\angle HKL = 30°$ and $LK = 6\sqrt{3}$, then $HL = 6$ and $HK = 12$. $MK = 6$ since the diagonals of a rectangle bisect each other.

17. $RS = 6$ and $RT = 6\sqrt{3} \approx 10.39$.

21. From vertex Z, draw an altitude to $\overline{XY}$; call the altitude $\overline{ZW}$. In the 30-60-90 $\triangle$, $WX = 6$ and $ZW = 6\sqrt{3}$. In the 45-45-90 $\triangle$, $\overline{YW} = 6\sqrt{3}$. $XY = YW + WX = 6\sqrt{3} + 6 \approx 16.39$.

25. 60°; $200^2 + x^2 = 400^2$

$x^2 = 400^2 - 200^2$

$x^2 = 120,000$

$x \approx 346$

The jogger travels $(200 + 346) - 400 = 146$ feet further.

29. Since $\triangle MNQ$ is equiangular and $\overline{NR}$ bisects $\angle MNQ$ and $\overline{QR}$ bisects $\angle MQN$, $m\angle RQN = 30° = m\angle RNQ$. From R, draw an altitude to $\overline{NQ}$. Name the altitude $\overline{RP}$. $NR = RQ = 6$. In 30-60-90 $\triangle RPQ$, $RP = 3$ and $PQ = 3\sqrt{3}$. NQ therefore equals $6\sqrt{3} \approx 10.39$.

33. Draw in altitude $\overline{CD}$. In right $\triangle CDB$, if $BC = 12$, then $CD = 6$ and $DB = 6\sqrt{3}$. In right $\triangle ACD$, if $CD = 6$, then $AD = 6$ and $AC = 6\sqrt{2}$. $AB = 6 + 6\sqrt{3} \approx 16.39$.

SECTION 5.6: Segments Divided Proportionally

1. Let $5x$ = the amount of ingredient A;
$4x$ = amount of ingredient B;
$6x$ = amount of ingredient C.
$$5x + 4x + 6x = 90$$
$$15x = 90$$
$$x = 6$$
30 ounces of ingredient A;
24 ounces of ingredient B;
36 ounces of ingredient C.

5. Let $EF = x$, $FG = y$, and $GH = z$
$AD = 5 + 4 + 3 = 12$ so that

$$\frac{AB}{AD} = \frac{EF}{EH} \qquad \frac{BC}{AD} = \frac{FG}{EH} \qquad \frac{CD}{AD} = \frac{GH}{EH}$$

$$\frac{5}{12} = \frac{x}{10} \qquad \frac{4}{12} = \frac{y}{10} \qquad \frac{3}{12} = \frac{z}{10}$$

$$12x = 50 \qquad 12y = 40 \qquad 12z = 30$$

$$x = \frac{50}{12} = 4\frac{1}{6} \quad y = \frac{40}{12} = 3\frac{1}{3} \quad z = \frac{30}{12} = 2\frac{1}{2}$$

$$EF = 4\frac{1}{6} \qquad FG = 3\frac{1}{3} \qquad GH = 2\frac{1}{2}$$

9. Let $EC = x$.
$$\frac{5}{12} = \frac{7}{x}$$
$$5x = 84$$
$$x = \frac{84}{5} \text{ or } 16\frac{4}{5}; \; EC = 16\frac{4}{5}$$

13. a. No

b. Yes

17. Let $NP = x = MQ$.
$$\frac{x}{12} = \frac{8}{x}$$
$$x^2 = 96$$
$$x = \sqrt{96} = 4\sqrt{6}$$
$$NP = 4\sqrt{6} \approx 9.80$$

21. If $RS = 6$ and $RT = 12$, then $\triangle RST$ is a
30-60-90 $\triangle$ and $ST = 6\sqrt{3}$. Let $SV = x$. Then
$VT = 6\sqrt{3} - x$.
$$\frac{6}{12} = \frac{x}{6\sqrt{3} - x}$$
$$36\sqrt{3} - 6x = 12x$$
$$36\sqrt{3} = 18x$$
$$x = 2\sqrt{3}$$
$$SV = 2\sqrt{3} \approx 3.46$$
$$VT = 6\sqrt{3} - 2\sqrt{3} = 4\sqrt{3} \approx 6.93$$

25. a. True

b. True

29. 1. Given

2. Means-Extremes Property

3. Addition Property of Equality

4. Distributive Property

5. Means-Extremes Property

6. Substitution

33. Given: $\triangle XYZ$; $\overline{YW}$ bisects $\angle XYZ$
$\overline{WX} \cong \overline{WZ}$
Prove: $\triangle XYZ$ is isosceles

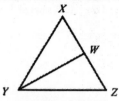

STATEMENTS	REASONS
1. $\triangle XYZ$; $\overline{YW}$ bisects $\angle XYZ$	1. Given
2. $\dfrac{YX}{YZ} = \dfrac{WX}{WZ}$	2. If a ray bisects one $\angle$ of a $\triangle$, then it divides the opposite side into segments that are proportional to the two sides which form that angle.
3. $\overline{WX} \cong \overline{WZ}$	3. Given
4. $WX = WZ$	4. If two segments are $\cong$, then their measures are equal. into four $\cong$ $\angle$s.
5. $\dfrac{YX}{YZ} = \dfrac{WX}{WX} = 1$	5. Substitution
6. $YX = YZ$	6. Means-Extremes Property
7. $\overline{YX} \cong \overline{YZ}$	7. If 2 segments are equal in measure, they are $\cong$.
8. $\triangle XYZ$ is isosceles	8. If 2 sides of a $\triangle$ are $\cong$, then the $\triangle$ is isosceles.

37. a. Let $CD = x$. Then $DB = 5 - x$.

$$\frac{CD}{CA} = \frac{DB}{BA}$$

$$\frac{x}{4} = \frac{5-x}{6}$$

$$4(5-x) = 6x$$

$$20 - 4x = 6x$$

$$10x = 20$$

$$x = 2$$

$$CD = 2$$

$$DB = 5 - 2 = 3$$

b. Let $CE = x$. Then $EA = 4 - x$.

$$\frac{CE}{BC} = \frac{EA}{BA}$$

$$\frac{x}{5} = \frac{4-x}{6}$$

$$5(4-x) = 6x$$

$$20 - 5x = 6x$$

$$11x = 20$$

$$x = \frac{20}{11}$$

$$CE = \frac{20}{11}$$

$$EA = 4 - \frac{20}{11} = \frac{24}{11}$$

c. Let $BF = x$. Then $FA = 6 - x$.

$$\frac{BF}{BC} = \frac{FA}{CA}$$

$$\frac{x}{5} = \frac{6-x}{4}$$

$$5(6-x) = 4x$$

$$30 - 5x = 4x$$

$$9x = 30$$

$$x = \frac{10}{3}$$

$$BF = \frac{10}{3}$$

$$FA = 6 - \frac{10}{3} = \frac{8}{3}$$

d. $\dfrac{BD}{DC} \cdot \dfrac{CE}{EA} \cdot \dfrac{AF}{FB} = \dfrac{3}{2} \cdot \dfrac{\frac{20}{11}}{\frac{24}{11}} \cdot \dfrac{\frac{8}{3}}{\frac{10}{3}} = \dfrac{3}{2} \cdot \dfrac{20}{24} \cdot \dfrac{8}{10} = 1$

CHAPTER REVIEW

1. False

2. True

3. False

4. True

5. True

6. False

7. True

8. a. $x^2 = 18$

$$x = \pm\sqrt{18} = \pm 3\sqrt{2} \approx \pm 4.24$$

b. $7(x-5) = 3(2x-3)$

$$7x - 35 = 6x - 9$$

$$x = 26$$

c. $6(x+2) = 2(x+4)$

$$6x + 12 = 2x + 8$$

$$4x = -4$$

$$x = -1$$

d. $7(x+3) = 5(x+5)$

$$7x + 21 = 5x + 25$$

$$2x = 4$$

$$x = 2$$

e. $(x-2)(x-1) = (x-5)(2x+1)$

$$x^2 - 3x + 2 = 2x^2 - 9x - 5$$

$$x^2 - 6x - 7 = 0$$

$$(x-7)(x+1) = 0$$

$$x = 7 \text{ or } x = -1$$

f. $\qquad 5x(x+5) = 9(4x+4)$

$$5x^2 + 25x = 36x + 36$$

$$5x^2 - 11x - 36 = 0$$

$$(5x+9)(x-4) = 0$$

$$x = -\frac{9}{5} \text{ or } x = 4$$

g. $(x-1)(3x-2) = 10(x+2)$

$$3x^2 - 5x + 2 = 10x + 20$$

$$3x^2 - 15x - 18 = 0$$

$$x^2 - 5x - 6 = 0$$

$$(x-6)(x+1) = 0$$

$$x = 6 \text{ or } x = -1$$

h. $(x+7)(x-2) = 2(x+2)$

$$x^2 + 5x - 14 = 2x + 4$$

$$x^2 + 3x - 18 = 0$$

$$(x+6)(x-3) = 0$$

$$x = -6 \text{ or } x = 3$$

9. Let x = cost of the six containers.

$$\frac{4}{2.52} = \frac{6}{x}$$

$$4x = 15.12$$

$$x = 3.78$$

The six containers cost $3.78.

10. Let x = the number of packages you can buy for $2.25.

$$\frac{2}{0.69} = \frac{x}{2.25}$$

$$0.69x = 4.50$$

$$x = \frac{450}{69} = 6\frac{12}{23}$$

With $2.25, you can buy 6 packages of M&M's.

11. Let x = cost of the rug that is 12 square meters.

$$\frac{20}{132} = \frac{12}{x}$$
$$20x = 1584$$
$$x = 79.20$$

The 12 square meters rug will cost $79.20.

12. Let the measure of the sides of the quadrilateral be $2x, 3x, 5x$ and $7x$.

$$2x + 3x + 5x + 7x = 68$$
$$17x = 68$$
$$x = 4$$

The length of the sides are 8, 12, 20 and 28.

13. Let the width of the similar rectangle be x.

$$\frac{18}{12} = \frac{27}{x}$$
$$\frac{3}{2} = \frac{27}{x}$$
$$3x = 54$$
$$x = 18$$

The width of the similar rectangle is 18.

14. Let x and y be the lengths of the other two sides.

$$\frac{6}{15} = \frac{8}{x} = \frac{9}{y}$$

$$\frac{2}{5} = \frac{8}{x} \quad \text{and} \quad \frac{2}{5} = \frac{9}{y}$$
$$2x = 40 \qquad\qquad 2y = 45$$
$$x = 20 \qquad\qquad y = \frac{45}{2} = 22\frac{1}{2}$$

The other two sides have lengths 20 and $22\frac{1}{2}$.

15. Let the measure of the angle be x; the measure of the supplement would be $180 - x$; the measure of the complement would be $90 - x$.

$$\frac{180 - x}{90 - x} = \frac{5}{2}$$
$$2(180 - x) = 5(90 - x)$$
$$360 - 2x = 450 - 5x$$
$$3x = 90$$
$$x = 30$$

The measure of the supplement is 150°.

16. **a.** SSS ~

b. AA

c. SAS ~

d. SSS ~

17. Given: $ABCD$ is a parallelogram;
$\overline{DB}$ intersects $\overline{AE}$ at pt. F

Prove: $\dfrac{AF}{EF} = \dfrac{AB}{DE}$

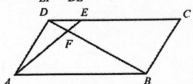

STATEMENTS	REASONS
1. $ABCD$ is a parallelogram $\overline{DB}$ intersects $\overline{AE}$ at pt. F	1. Given
2. $\overline{DC} \parallel \overline{AB}$	2. Opp. sides of a parallelogram are $\parallel$.
3. $\angle CDB \cong \angle ABD$	3. If 2 $\parallel$ lines are cut by a trans., then the alt. int. $\angle$s $\cong$.
4. $\angle DEF \cong \angle BAF$	4. Same as (3).
5. $\triangle DFE \sim \triangle BFA$	5. AA
6. $\dfrac{AF}{EF} = \dfrac{AB}{DE}$	6. Corresp. sides of ~ $\triangle$s are proportional.

18. Given: $\angle 1 \cong \angle 2$

Prove: $\dfrac{AB}{AC} = \dfrac{BE}{CD}$

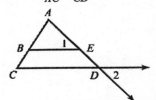

STATEMENTS	REASONS
1. $\angle 1 \cong \angle 2$	1. Given
2. $\angle ADC = \angle 2$	2. If 2 lines intersect, then the vertical formed are $\cong$.
3. $\angle ADC \cong \angle 1$	3. Transitive Prop. for Congruence.
4. $\angle A \cong \angle A$	4. Identity.
5. $\triangle BAE \sim \triangle CAD$	5. AA
6. $\dfrac{AB}{AC} = \dfrac{BE}{CD}$	6. Corresp. sides of ~ $\triangle$s are proportional.

19. Since the $\triangle$ s are $\sim$, $m\angle A = m\angle D$.

$50 = 2x + 40$

$10 = 2x$

$x = 5$

$m\angle D = 2(5) + 40 = 50°$

$m\angle E = 33°$

$m\angle F = 180 - 33 - 50 = 97°$

20. With $\angle B \cong \angle F$, and $\angle C \cong \angle E$,

$\triangle ABC \sim \triangle DFE$. It follows that

$$\frac{AB}{DF} = \frac{AC}{DE} = \frac{BC}{FE}$$

Substituting in gives

$$\frac{AB}{2} = \frac{9}{3} \text{ or } \frac{3}{1} \text{ and } \frac{9}{3} \text{ or } \frac{3}{1} = \frac{BC}{4}$$

$AB = 6$ and $BC = 12$.

21. $\frac{BD}{AD} = \frac{BE}{EC}$; let $AB = x$

$\frac{6}{x} = \frac{8}{4} \text{ or } \frac{2}{1}$

$2x = 6$

$x = 3$

22. $\frac{BD}{BA} = \frac{DE}{AC}$; let $AC = x$; $BA = 12$

$\frac{8}{12} = \frac{2}{3} = \frac{3}{x}$

$2x = 9$

$x = 4\frac{1}{2}$

$AC = 4\frac{1}{2}$

23. $\frac{BD}{BA} = \frac{BE}{BC}$; let $BC = x$; $BD = 8$

$\frac{8}{10} \text{ or } \frac{4}{5} = \frac{5}{x}$

$4x = 25$

$x = 6\frac{1}{4}$

$BC = 6\frac{1}{4}$

24. Since $\overline{GJ}$ bisects $\angle FGH$, we can write the

proportion $\frac{FJ}{FG} = \frac{JH}{GH}$; let $JH = x$.

$\frac{7}{10} = \frac{x}{8}$

$10x = 56$

$x = \frac{56}{10} = 5\frac{3}{5}$

$JH = 5\frac{3}{5}$

25. Since $\overline{GJ}$ bisects $\angle FGH$, and $GF:GH = 1:2$,

we can write the proportion $\frac{GF}{GH} = \frac{FJ}{JH}$; let

$JH = x$.

$\frac{1}{2} = \frac{5}{x}$

$x = 10$

$JH = 10$

26. Since $\overline{GJ}$ bisects $\angle FGH$, we can write the

proportion $\frac{FG}{GH} = \frac{FJ}{JH}$; let $FJ = x$ and

$JH = 15 - x$.

$\frac{8}{12} = \frac{x}{15-x}$

$\frac{2}{3} = \frac{x}{15-x}$

$2(15 - x) = 3x$

$30 - 2x = 3x$

$30 = 5x$

$x = 6$

$FJ = 6$

27. Let $MK = x$, then $\frac{MK}{HJ} = \frac{EM}{FH}$.

$\frac{x}{5} = \frac{6}{10} \text{ or } \frac{3}{5}$

$x = 3$

$MK = 3$

Let $EO = y$ and $OM = 6 - y$.

Then $\frac{y}{2} = \frac{6-y}{8}$

$8y = 2(6 - y)$

$8y = 12 - 2y$

$10y = 12$

$y = \frac{12}{10} = 1\frac{1}{5}$

$EO = 1\frac{1}{5}$

$EK = 9$

28. If a line bisects one side of a triangle and is parallel to a second side, then it bisects the third side.

Given: $\overline{DE}$ bisects $\overline{AC}$

        $\overline{DE} \parallel \overline{BC}$

Prove: $\overline{DE}$ bisects $\overline{AB}$

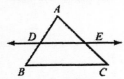

STATEMENTS	REASONS
1. $\overline{DE}$ bisects $\overline{AC}$	**1.** Given
2. $AE = EC$	**2.** Bisecting a segment forms 2 segments of equal measure.
3. $\overline{DE} \parallel \overline{BC}$	**3.** Given
4. $\dfrac{AD}{DB} = \dfrac{AE}{EC}$	**4.** If a line is parallel to one side of a triangle and intersects the other sides, then it divides these sides proportionally.
5. $AD \cdot EC = DB \cdot AE$	**5.** Means-Extremes Prop.
6. $AD = DB$	**6.** Division Prop. of Eq.
7. $\overline{DE}$ bisects $\overline{AB}$	**7.** If a segment has been divided into 2 segments of equal measure, the segment has been bisected.

29. The diagonals of a trapezoid divide themselves proportionally.

Given: $ABCD$ is a trapezoid

     with $\overline{BC} \parallel \overline{AD}$ and

     diagonals $\overline{BD}$ and $\overline{AC}$

Prove: $\dfrac{BE}{ED} = \dfrac{EC}{AE}$

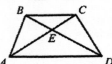

STATEMENTS	REASONS
1. $ABCD$ is a trapezoid with $\overline{BC} \parallel \overline{AD}$ and diagonals $\overline{BD}$ and $\overline{AC}$.	**1.** Given
2. $\angle CBE \cong \angle ADE$ and $\angle BCE \cong \angle DAE$	**2.** If 2 $\parallel$ lines are cut by a trans., then the alt. int. $\angle$s are $\cong$.
3. $\triangle BCE \cong \triangle DAE$	**3.** AA
4. $\dfrac{BE}{ED} = \dfrac{EC}{AE}$	**4.** Corresponding sides of $\sim$ $\triangle$s are proportional.

30. a. $\dfrac{BD}{AD} = \dfrac{AD}{DC}$

Let $DC = x$.

$$\dfrac{3}{5} = \dfrac{5}{x}$$

$$3x = 25$$

$$x = \dfrac{25}{3}$$

$$DC = 8\dfrac{1}{3}$$

b. $\dfrac{DC}{AC} = \dfrac{AC}{BC}$

Let $BD = x$ and $BC = x + 4$

$$\dfrac{4}{5} \text{ or } \dfrac{2}{3} = \dfrac{10}{x+4}$$

$$2(x+4) = 50$$

$$2x + 8 = 50$$

$$2x = 42$$

$$x = 21$$

$$BD = 21$$

c. $\dfrac{BD}{BA}=\dfrac{BA}{BC}$

Let $BA=x$.

$\dfrac{2}{x}=\dfrac{x}{6}$

$x^2=12$

$x=\sqrt{12}=2\sqrt{3}$

$BA=2\sqrt{3}\approx 3.46$

d. $\dfrac{DA}{AC}=\dfrac{AC}{BC}$

Let $DC=x$ and $BC=x+3$.

$\dfrac{x}{3\sqrt{2}}=\dfrac{3\sqrt{2}}{x+3}$

$x(x+3)=18$

$x^2+3x-18=0$

$(x+6)(x-3)=0$

$x=-6$ or $x=3$; reject $x=-6$

$DC=3$.

31. a. $\dfrac{AD}{BD}=\dfrac{BD}{DC}$

Let $DC=x$.

$\dfrac{9}{12}$ or $\dfrac{3}{4}=\dfrac{12}{x}$

$3x=48$

$x=16$

$DC=16$

b. $\dfrac{DC}{BC}=\dfrac{BC}{AC}$

Let $AD=x$ and $AC=x+5$.

$\dfrac{5}{15}$ or $\dfrac{1}{3}=\dfrac{15}{x+5}$

$x+5=45$

$x=40$

$AD=40$

c. $\dfrac{AD}{AB}=\dfrac{AB}{AC}$

Let $AB=x$ and $AC=10$.

$\dfrac{2}{x}=\dfrac{x}{10}$

$x^2=20$

$x=\sqrt{20}=2\sqrt{5}$

$AB=2\sqrt{5}\approx 4.47$

d. $\dfrac{AD}{AB}=\dfrac{AB}{AC}$

Let $AD=x$ and $AC=x+2$.

$\dfrac{x}{2\sqrt{6}}=\dfrac{2\sqrt{6}}{x+2}$

$x(x+2)=24$

$x^2+2x-24=0$

$(x+6)(x-4)=0$

$x=-6$ or $x=4$; reject $x=-6$

$AD=4$.

32. a. $x=30$. Since the leg is half of the hypotenuse, the angle opposite the leg must be 30°.

b. Half of the base is 10. In the right $\triangle$, 1 side has length 10 and the hypotenuse has length 26. The other side has length 24 since (10, 24, 26) is a multiple of (5, 12, 13).

c. $x^2=12^2+16^2$

$x^2=144+256$

$x^2=400$

$x=20$

or (12, 16, 20) is a multiple of (3, 4, 5).

d. The unknown length of the right $\triangle$ is 8 using the Triple (8, 15, 17).

$x=16$.

33. In rect. $ABCD$, $BC=24$ and since E is a midpoint, $BE=12$ and $EC=12$. $CD=16$ and $FD=7$. There are three right triangles for which the Pythagorean Triples apply. $AE=20$ using (12, 16, 20) which is a multiple of (3, 4, 5). $EF=15$ using (9, 12, 15) which is also a multiple of (3, 4, 5). $AF=25$ using the Triple (7, 24, 25).

34. In a square there are two 45-45-90 $\triangle$ s. If the length of the side of the square is 4 inches, then the length of the diagonal is $4\sqrt{2}\approx 5.66$ inches.

35. In a square there are two 45°-45°-90° $\triangle$ s. If the length of the diagonal is 6, then $6=a\sqrt{2}$. Solving for a gives

$a=\dfrac{6}{\sqrt{2}}=\dfrac{6\sqrt{2}}{2}=3\sqrt{2}$

Hence, the length of a side is $3\sqrt{2}\approx 4.24$ cm.

36. Since the diagonals of a rhombus are perpendicular and bisect each other, there are 4 right triangles formed whose sides are the lengths 24 cm and 7 cm. The hypotenuse must then have a length of 25. Since they hypotenuse of a right triangle is the side of the rhombus, the side has length 25 cm.

37. The altitude to one side of an equilateral triangle divides it into two 30°-60°-90° $\triangle$ s.

The altitude is the side opposite the 60 degree angle and is equal in length to one-half the length of the hypotenuse times $\sqrt{3}$. Hence, the altitude has length $5\sqrt{3}\approx 8.66$ in.

38. The altitude to one side of an equilateral triangle divides it into two 30°-60°-90° △ .

The altitude is the side opposite the 60 degree angle and is equal in length to one-half the length of the hypotenuse times $\sqrt{3}$. If H represents the length of the hypotenuse, then

$$6 = \frac{1}{2} \cdot H \cdot \sqrt{3}$$
$$12 = H \cdot \sqrt{3}$$
$$H = \frac{12}{\sqrt{3}} \cdot \frac{\sqrt{3}}{\sqrt{3}} = \frac{12\sqrt{3}}{3} = 4\sqrt{3}$$

The length of the sides of the △ is $4\sqrt{3} \approx 6.93$ in.

39. The altitude to the side of length 14 separates it into two parts the lengths of these are given by x and $14 - x$.

If the length of the altitude is H, we can use the Pythagorean Theorem on the two right triangles to get $x^2 + H^2 = 13^2$ and $(14 - x)^2 + H^2 = 15^2$. Subtracting the first equation from the second, we have

$$196 - 29x + x^2 + H^2 = 225$$
$$\underline{x^2 + H^2 = 169}$$
$$196 - 28x = 56$$
$$-28x = -140$$
$$x = 5$$

Now we use x to find H.
$x^2 + H^2 = 13^2$ becomes
$$5^2 + H^2 = 169$$
$$H^2 = 144$$
$$H = 12$$
The length of the altitude is 12 cm.

40. a. Let the length of the hypotenuse common to both △ s be H.

$$\frac{1}{2} \cdot H \cdot \sqrt{3} = 9\sqrt{3}$$
$$H \cdot \sqrt{3} = 18\sqrt{3}$$
$$H = 18$$
$$y = \frac{1}{2} \cdot 18 = 9$$
$$x = 9\sqrt{2} \approx 12.73$$

b. $y = 6$ using the Triple (6, 8, 10). Since the length of the altitude to the hypotenuse is 6, 6 is the geometric mean for x and 8. That is,

$$\frac{x}{6} = \frac{6}{8}$$
$$8x = 36$$
$$x = \frac{36}{8} = \frac{9}{2} = 4\frac{1}{4}.$$

c. 6 is the geometric mean for y and x. But since $x = y + 9$, we have

$$\frac{y}{6} = \frac{6}{y+9}$$
$$y(y+9) = 36$$
$$y^2 + 9y - 36 = 0$$
$$(y+12)(y-3) = 0$$
$$y = -12 \text{ or } y = 3; \text{ reject } y = -12$$
$$\therefore y = 3 \text{ and } x = 12.$$

d. $4^2 + x^2 = \left(6\sqrt{2}\right)^2$
$$16 + x^2 = 72$$
$$x^2 = 56$$
$$x = \sqrt{56} = 2\sqrt{14} \approx 7.48$$
$$y = 13$$

41.

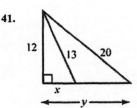

$x = 5$ using the Triple (5, 12, 13).
$y = 16$ using the Triple (12, 16, 20).
The ships are 11 km apart.

42. a. $14^2 < 12^2 + 13^2 \therefore$ acute $\triangle$

b. $11 + 5 \not> 18 \therefore$ no $\triangle$

c. $18^2 > 9^2 + 15^2 \therefore$ obtuse $\triangle$

d. $10^2 = 6^2 + 8^2 \therefore$ right $\triangle$

e. $8 + 7 \not> 16 \therefore$ no $\triangle$

f. $8^2 < 7^2 + 6^2 \therefore$ acute $\triangle$

g. $13^2 > 8^2 + 9^2 \therefore$ obtuse $\triangle$

h. $4^2 > 2^2 + 3^2 \therefore$ obtuse $\triangle$

CHAPTER TEST

1. a. $3{:}5$ or $\dfrac{3}{5}$

b. $\dfrac{25 \text{ mi}}{\text{gal}}$

2. a. $\dfrac{x}{5} = \dfrac{8}{13}$
$13x = 40$
$x = \dfrac{40}{13}$

b. $\dfrac{x+1}{5} = \dfrac{16}{x-1}$
$(x+1)(x-1) = 80$
$x^2 - 1 = 80$
$x^2 = 81$
$x = 9 \text{ or } -9$

3. $15°$ and $75°$

4. a. $m\angle W \cong m\angle T$. Let $m\angle T = x$
$m\angle T + m\angle R + m\angle S = 180°$
$x + 67 + 21 = 180$
$x = 92$
$m\angle W = 92°$

b. $\dfrac{UW}{RT} = \dfrac{WV}{TS}$
$\dfrac{6}{4} = \dfrac{x}{8}$
$48 = 4x$
$x = 12$
$WV = 12$

5. a. SAS~

b. AA

6. $\triangle ABC \sim \triangle ACD \sim \triangle CBD$

7. a. $c = \sqrt{a^2 + b^2}$
$c = \sqrt{5^2 + 4^2}$
$c = \sqrt{25 + 16}$
$c = \sqrt{41}$

b. $a^2 + b^2 = c^2$
$a^2 = c^2 - b^2$
$a = \sqrt{c^2 - b^2}$
$a = \sqrt{8^2 - 6^2}$
$a = \sqrt{64 - 36}$
$a = \sqrt{28} = 2\sqrt{7}$

8. a. $a^2 + b^2 = 15^2 + 8^2 = 225 + 64 = 289$
$c^2 = 17^2 = 289$
Yes

b. $a^2 + b^2 = 11^2 + 8^2 = 121 + 64 = 185$
$c^2 = 15^2 = 225$
No

9. $(AC)^2 = (AB)^2 + (BC)^2$
$AC = \sqrt{4^2 + 3^2}$
$AC = \sqrt{16 + 9}$
$AC = \sqrt{25} = 5$
$(DA)^2 = (AC)^2 + (DC)^2$
$DA = \sqrt{5^2 + 8^2}$
$DA = \sqrt{25 + 64}$
$DA = \sqrt{89}$

10. a. In the 45-45-90 $\triangle$, $XZ = 10$ and $YZ = 10$ so $XY = 10\sqrt{2}$ in.

b. In the 45-45-90 $\triangle$, $XY = 8\sqrt{2}$ so $XZ = 8$ cm.

11. a. In the 30-60-90 $\triangle$, $EF = 10$ so $DE = 5$ m.

b. In the 30-60-90 $\triangle$, $DF = 6\sqrt{3}$ so $EF = 12$ ft.

12. Let $EC = x$. Then $AC = 9 + x$.
$\dfrac{AB}{AD} = \dfrac{AC}{AE}$
$\dfrac{6+8}{6} = \dfrac{9+x}{9}$
$6(9 + x) = 126$
$54 + 6x = 126$
$6x = 72$
$x = 12$
$EC = 12$

13. Let $PQ = x$. Then $QM = 10 - x$.

$$\frac{PQ}{PN} = \frac{QM}{MN}$$

$$\frac{x}{6} = \frac{10 - x}{9}$$

$$9x = 6(10 - x)$$

$$9x = 60 - 6x$$

$$15x = 60$$

$$x = 4$$

$$PQ = 4$$

$$QM = 10 - 4 = 6$$

14. 1

15. **S1.** $\overline{MN} \parallel \overline{QR}$

   R1. Given

   R2. Corresponding $\angle$ s are $\cong$.

   S3. $\angle P \cong \angle P$

   R4. AA

16. **R1.** Given

   R2. Identity

   R3. Given

   R5. Substitution

   R6. SAS~

   S7. $\angle PRC \cong \angle B$

Chapter 6: Circles

SECTION 6.1: Circles and Related Segments and Angles

1. 29°

5. 56.6°

9. **a.** 90°
   b. 270°
   c. 135°
   d. 135°

13. **a.** 72°
    b. 144°
    c. 36°
    d. 72°
    e. Draw in $\overline{OA}$. In $\triangle BOA$, m$\angle BOA = 144°$; therefore the m$\angle ABO = 18°$

17. $RV = 4$ and let $RQ = x$.
    Using the Pythagorean Theorem, we have
    $$x^2 + 4^2 = (x+2)^2$$
    $$x^2 + 16 = x^2 + 4x + 4$$
    $$12 = 4x$$
    $$x = 3$$
    $$RQ = 3$$

21. 90°; Square

25. **a.**

At 6:30 PM, the hour hand is half the distance from 6 to the 7. Therefore, the angle measure is 15°.

   b.

At 5:40 AM the hour hand is $\frac{2}{3}$ the distance from 5 to 6. Therefore, the angle measure is found by adding 60 and 10. The angle is 70°.

29. 45°

33. Proof: Using the chords $\overline{AB}$, $\overline{BC}$, $\overline{CD}$ and $\overline{AD}$ in $\odot O$ as sides of inscribed angles, $\angle B \cong \angle D$ and $\angle A \cong \angle C$ since they are inscribed angles intercepting the same arc. $\triangle ABE \sim \triangle CED$ by AA.

37. Prove: If two inscribed angles intercept the same arc, then these angles are congruent.
    Given: A circle with inscribed angles A and D intercepting $\overset{\frown}{BC}$.
    Prove: $\angle A \cong \angle D$

Proof: Since both inscribed angles, A and D, intercept $\overset{\frown}{BC}$, m$\angle A = \frac{1}{2}$m$\overset{\frown}{BC}$ and

m$\angle D = \frac{1}{2}$m$\overset{\frown}{BC}$. Therefore, m$\angle A = m\angle D$ which means $\angle A \cong \angle D$.

41. Given: $\odot O$ with inscribed $\angle RSW$ and diameter $\overline{ST}$
    Prove: m$\angle RSW = \frac{1}{2} \cdot m\overset{\frown}{RW}$

Proof: In $\odot O$, diameter $\overline{ST}$ is one side of inscribed $\angle RST$ and one side of inscribed $\angle TSW$. Using Case (1), m$\angle RST = \frac{1}{2} \cdot m\overset{\frown}{RT}$ and

m$\angle TSW = \frac{1}{2} \cdot m\overset{\frown}{TW}$. By the Addition Property of Equality,
m$\angle RST + m\angle TSW = \frac{1}{2} \cdot m\overset{\frown}{RT} + \frac{1}{2} \cdot m\overset{\frown}{TW}$. But
since m$\angle RSW = m\angle RST + m\angle TSW$, we have
m$\angle RSW = \frac{1}{2} \cdot m\overset{\frown}{RT} + \frac{1}{2} \cdot m\overset{\frown}{TW}$. Factoring the $\frac{1}{2}$

out, we have m$\angle RSW = \frac{1}{2}\left(m\overset{\frown}{RT} + m\overset{\frown}{TW}\right)$. But

m$\overset{\frown}{RT} + m\overset{\frown}{TW} = m\overset{\frown}{RW}$. Therefore
m$\angle RSW = \frac{1}{2} \cdot m\overset{\frown}{RW}$.

SECTION 6.2: More Angle Relationships in the Circle

1. If $m\widehat{AB} = 92°$, $m\widehat{DA} = 114°$, and $m\widehat{BC} = 138°$, then $m\widehat{DC} = 16°$.

   a. $m\angle 1 = \frac{1}{2}(16) = 8°$

   b. $m\angle 2 = \frac{1}{2}(92) = 46°$

   c. $m\angle 3 = \frac{1}{2}(92-16) = \frac{1}{2}(76) = 38°$

   d. $m\angle 4 = \frac{1}{2}(92+16) = \frac{1}{2}(108) = 54°$

   e. $m\angle 5 = 180 - 54 = 126°$ or
   $m\angle 5 = \frac{1}{2}(114+138) = \frac{1}{2}(252) = 126°$

5. Let $m\widehat{RT} = x$, then $m\widehat{TS} = 4x$
   $x + 4x = 180$
   $5x = 180$
   $x = 36$
   $m\widehat{RT} = 36°$
   $m\angle RST = \frac{1}{2}(m\widehat{RT})$
   $m\angle RST = \frac{1}{2}(36)$
   $m\angle RST = 18°$

9. a. $m\angle ACB = \frac{1}{2}(m\widehat{BC})$ or $68 = \frac{1}{2}(m\widehat{BC})$ or
   $m\widehat{BC} = 136°$.

   b. $m\widehat{BDC} = 360 - 136 = 224°$

   c. $m\angle ABC = \frac{1}{2}(136) = 68°$

   d. $m\angle A = \frac{1}{2}(224-136) = \frac{1}{2}(88) = 44°$

13. a. $m\widehat{RT} = 120°$

   b. $m\widehat{RST} = 240°$

   c. $m\angle 3 = \frac{1}{2}(240-120) = \frac{1}{2}(120) = 60°$

17. Let $m\widehat{CE} = x$ and $m\widehat{BD} = y$.
   $62 = \frac{1}{2}(x+y)$
   $26 = \frac{1}{2}(x-y)$

   $124 = x+y$
   $\underline{52 = x-y}$
   $176 = 2x$
   $88 = x$
   $m\widehat{CE} = 88°$
   $y = 36$
   $m\widehat{BD} = 36°$

21. 1. $\overline{AB}$ and $\overline{AC}$ are tangents to $\odot O$ from A

   2. Measure of an $\angle$ formed by a tangent and a chord $= \frac{1}{2}$ the arc measure.

   3. Substitution

   4. If 2 $\angle$s are = in measure, they are $\cong$.

   5. $\overline{AB} \cong \overline{AC}$

   6. $\triangle ABC$ is isosceles

25. If $m\widehat{AB} = x$, then $m\widehat{ADB} = 360 - x$. Then
   $m\angle 1 = \frac{1}{2}(m\widehat{ADB} - m\widehat{AB})$
   $= \frac{1}{2}(360 - x - x)$
   $= \frac{1}{2}(360 - 2x)$
   $= 180° - x$

29. Each arc of the circle is 72°.
   $m\angle 1 = \frac{1}{2}(72) = 36°$
   $m\angle 2 = \frac{1}{2}(144+72) = \frac{1}{2}(216) = 108°$

33. $\angle x \cong \angle x$; $\angle R \cong \angle W$; also, $\angle RVW \cong \angle WSX$

37. If 2 parallel lines intersect a circle, then the intercepted arcs between these lines are congruent.

Given: $\overline{BC} \parallel \overline{AD}$

Prove: $\overparen{AB} \cong \overparen{CD}$

Proof: Draw $\overline{AC}$. If $\overline{BC} \parallel \overline{AD}$, then $\angle 1 \cong \angle 2$ or $m\angle 1 = m\angle 2$. $m\angle 1 = \frac{1}{2} m\overparen{AB}$ and $m\angle 2 = \frac{1}{2} m\overparen{CD}$. Therefore, $\frac{1}{2} m\overparen{AB} = \frac{1}{2} m\overparen{CD}$ or $m\overparen{AB} = m\overparen{CD}$. $\overparen{AB} \cong \overparen{CD}$ since they are in the same circle and have equal measures.

41. If one side of an inscribed triangle is a diameter, then the triangle is a right triangle.

Given: $\triangle ACB$ inscribed in $\odot O$ with $\overline{AB}$ a diameter

Prove: $\triangle ABC$ is a right $\triangle$

Proof: If $\triangle ACB$ is inscribed in $\odot O$ and $\overline{AB}$ is a diameter, then $\angle ACB$ must be a right angle because an angle inscribed in a semicircle is a right angle. $\triangle ACB$ is a right triangle since it contains a right angle.

45. Given: Quad. *RSTV* inscribed in $\odot Q$

Prove: $m\angle R + m\angle T = m\angle V + m\angle S$

Proof: $m\angle R = \frac{1}{2} m\overparen{VTS}$ and $m\angle T = \frac{1}{2} m\overparen{VRS}$

$\therefore m\angle R + m\angle T = \frac{1}{2} m\overparen{VTS} + \frac{1}{2} m\overparen{VRS}$

$\qquad = \frac{1}{2}\left(m\overparen{VTS} + m\overparen{VRS}\right)$

$\qquad = \frac{1}{2}(360°)$

$\qquad = 180°$

Similarly,

$m\angle V + m\angle S = \frac{1}{2} m\overparen{RST} + \frac{1}{2} m\overparen{RVT}$

$\qquad = \frac{1}{2}\left(m\overparen{RST} + m\overparen{RVT}\right)$

$\qquad = \frac{1}{2}(360°)$

$\qquad = 180°$

$\therefore m\angle R + m\angle T = m\angle V + m\angle S$

SECTION 6.3: Line and Segment Relationships in the Circle

1. $\triangle OCD$ is an equilateral triangle with $m\angle COD = 60°$. Since $\overline{OE} \perp \overline{CD}$, $\overline{OF}$ bisects $\overline{CD}$. $m\overparen{CF} = m\overparen{FD}$ and $m\angle COF = m\angle FOD = 30°$. $m\overparen{CF}$ must equal 30°.

5. a.

b.

c.

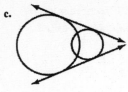

d.

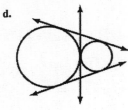

e.

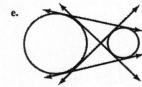

9. Let $DE = x$ and $EC = 16 - x$. Then
$$8 \cdot 6 = x(16 - x)$$
$$48 = 16x - x^2$$
$$x^2 - 16x - 48 = 0$$
$$(x - 4)(x - 12) = 0$$
$$x = 4 \text{ or } x = 12$$
$DE = 4$ and $EC = 12$ OR
$DE = 12$ and $EC = 4$

13. Let $EC = x$, then $DE = 2x$.
$$9 \cdot 8 = 2x \cdot x$$
$$72 = 2x^2$$
$$36 = x^2$$
$$x^2 - 36 = 0$$
$$(x - 6)(x + 6) = 0$$
$x = 6$ or $x = -6$; reject $x = -6$
$EC = 6$
$DE = 12$

17. If $AB = 4$ and $BC = 5$, then $AC = 9$.
Let $DE = x$ and $AE = x + 3$.
$$4 \cdot 9 = 3(x + 3)$$
$$36 = 3x + 9$$
$$27 = 3x$$
$$x = 9$$
$$DE = 9$$

21. Let $RS = TV = x$ and $RV = 6 + x$.
$$x^2 = 6(6 + x)$$
$$x^2 = 36 + 6x$$
$$x^2 - 6x - 36 = 0$$
Since the quadratic won't factor, we must use the quadratic formula.
$$x = \frac{-b \pm \sqrt{b^2 - 4ac}}{2a}$$
$$x = \frac{6 \pm \sqrt{36 - 4(1)(-36)}}{2(1)}$$
$$x = \frac{6 \pm \sqrt{36 + 144}}{2}$$
$$x = \frac{6 \pm \sqrt{180}}{2}$$
$$x = \frac{6 \pm 6\sqrt{5}}{2}$$
$$x = 3 \pm 3\sqrt{5}$$
$$RS = 3 \pm 3\sqrt{5}$$

25. If $\overline{AF}$ is tangent to $\odot O$ and $\overline{AC}$ is a secant to $\odot O$, then $(AF)^2 = AC \cdot AB$. If $\overline{AF}$ is a tangent to $\odot Q$ and $\overline{AE}$ is a secant to $\odot Q$, then $(AF)^2 = AE \cdot AD$. By substitution, $AC \cdot AB = AE \cdot AD$.

29. Yes; $\overline{AE} \cong \overline{CE}$; $\overline{DE} \cong \overline{EB}$

33. Let $AM = x$, then $MB = 14 - x$.
Let $BN = y$, then $NC = 16 - y$.
Let $PC = z$, then $AP = 12 - z$.
If tangent segments to a circle from an external point are congruent, $AM = AP$, $BN = MB$, and $PC = NC$ or
$$\begin{cases} x = 12 - z \\ y = 14 - x \\ z = 16 - y \end{cases} \text{ or } \begin{cases} x + z = 12 \\ x + y = 14 \\ y + z = 16 \end{cases}$$
Subtracting the first equation from the 2nd equation and using the 3rd equation, we have
$$y - z = 2$$
$$y + z = 16$$
Adding gives $2y = 18$ or $y = 9$. Solving for x and z we have $x = 5$ and $z = 7$. Therefore, $AM = 5$; $PC = 7$ and $BN = 9$.

37. $OA = 2$; $BP = 3$; $OD = x$; $DP = 10 - x$.
$$\frac{2}{x} = \frac{3}{10 - x} \text{ or } 20 - 2x = 3x. \text{ Solve to get } x = 4.$$
$\therefore OD = 4$; $DP = 6$; $AD = 2\sqrt{3}$ and $DB = 3\sqrt{3}$
so $AB = 5\sqrt{3}$ or $AB \approx 8.7$ in.

41. Let *x* represent the angle measure of the larger gear. It is intuitively obvious that the

$$\frac{\text{number of teeth in the larger gear}}{\text{number of teeth in the smaller gear}} = \frac{\text{angle measure in smaller gear}}{\text{angle measure in larger gear}}.$$

The proportion becomes

$$\frac{2}{1} = \frac{90}{x}$$
$$2x = 90$$
$$x = 45°$$

45. Given: $\overline{TX}$ is a secant segment and $\overline{TV}$ is a tangent at *V*

Prove: $(TV)^2 = TW \cdot TX$

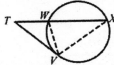

Proof: With secant $\overline{TX}$ and tangent $\overline{TV}$, draw in $\overline{WV}$ and $\overline{VX}$. $m\angle X = \frac{1}{2}m\widehat{WV}$ since $\angle X$ is an inscribed angle. $m\angle TVW = \frac{1}{2}m\widehat{WV}$ because it is formed by a tangent and a chord. By substitution, $m\angle TVW = m\angle X$ or $\angle TVW \cong \angle X$. $\angle T \cong \angle T$ and $\triangle TVW \sim \triangle TXV$. It follows that $\frac{TV}{TW} = \frac{TX}{TV}$ or

$$(TV)^2 = TW \cdot TX.$$

SECTION 6.4: Some Constructions and Inequalities for the Circle

1. $m\angle CQD < m\angle AQB$

5. $CD < AB$

9.

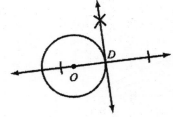

13.

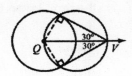

The measure of the angle formed by the tangents at *V* is 60°.

17. a. $\overline{OT}$

   b. $\overline{OD}$

21. Obtuse

25. a. $m\angle \widehat{AB} > m\angle \widehat{BC}$

   b. $AB > BC$

29. a. $\angle B$

   b. $\overline{AC}$

33.

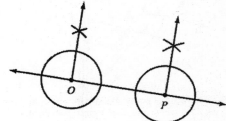

37. Draw in $\overline{OB}$ to form rt. $\triangle OMB$. Since $MB = 12$ and $OB = 13$, $OM = 5$. Draw in $\overline{OD}$ to form rt. $\triangle OND$. Since $ND = 5$ and $OD = 13$, $ON = 12$.

SECTION 6.5: Locus of Points

1. A, C, E

5.

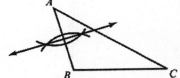

9.

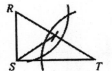

13. The locus of points at a distance of 3 inches from point O is a circle.

17. The locus of the midpoints of the chords in $\odot Q$ parallel to diameter $\overline{PR}$ is the perpendicular bisector of $\overline{PR}$.

21. The locus of points 1 cm from line p and 2 cm from line q are the four points A, B, C, and D as shown.

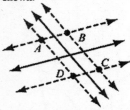

25. The locus of points at a distance of 2 cm from a sphere whose radius is 5 cm is two concentric spheres with the same center. The radius of one sphere is 3 cm and the radius of the other sphere is 7 cm.

29. The locus of points equidistant from an 8 foot ceiling and the floor is a parallel plane in the middle.

33.

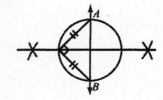

37.

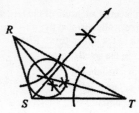

41.

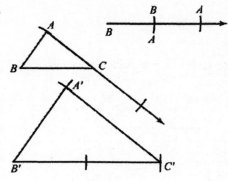

SECTION 6.6: Concurrence of Lines

1. Yes

5. Circumcenter

9. No

13. Midpoint of the hypotenuse

17.

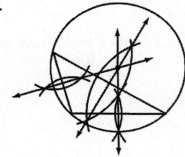

21.

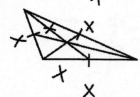

25. Using the figure as marked, we know that in the 30°-60°-90° triangle, the side opposite the 60 degree angle is $x\sqrt{3}$. Therefore, $x\sqrt{3} = 5$ or

$$x = \frac{5}{\sqrt{3}} \cdot \frac{\sqrt{3}}{\sqrt{3}} = \frac{5\sqrt{3}}{3}.$$

The radius would be $2 \cdot \frac{5\sqrt{3}}{3}$ or $\frac{10\sqrt{3}}{3}$.

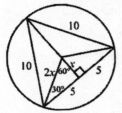

29. a. 4

   b. 6

   c. 10.5

33. Equilateral

37. a. Yes

   b. No

CHAPTER REVIEW

1. (9, 12, 15) is a multiple of the Pythagorean Triple (3, 4, 5). Therefore, the distance from the center of the circle to the chord is 9 mm.

2. (8, 15, 17) is a Pythagorean Triple. Therefore, the length of half of the chord is 15 and the length of the chord is 30 cm.

3. $r^2 = 5^2 + 4^2$
   $r^2 = 25 + 16$
   $r^2 = 41$
    $r = \sqrt{41}$

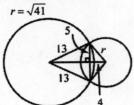

4. The radius of each circle has a length of $6\sqrt{2}$.

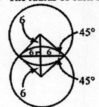

5. $m\angle B = \frac{1}{2}\left(m\widehat{AD} - m\widehat{AC}\right)$

    $25 = \frac{1}{2}\left(140 - m\widehat{AC}\right)$

    $50 = 140 - m\widehat{AC}$

   $m\widehat{AC} = 90°$

   $m\widehat{DC} = 360 - (140 + 90) = 130°$

6. $m\widehat{AC} = 360 - (190 + 120) = 50°$

   $m\angle B = \frac{1}{2}\left(m\widehat{AD} - m\widehat{AC}\right)$

   $m\angle B = \frac{1}{2}(120 - 50) = \frac{1}{2}(70) = 35°$

7. If $m\angle EAD = 70°$ then $m\widehat{AD} = 140°$.

   $m\angle B = \frac{1}{2}\left(m\widehat{AD} - m\widehat{AC}\right)$

    $30 = \frac{1}{2}\left(140 - m\widehat{AC}\right)$

    $60 = 140 - m\widehat{AC}$

   $m\widehat{AC} = 80°$

8. If $m\angle D = 40°$ then $m\widehat{AC} = 80°$.

   $m\widehat{AD} = 360 - (80 + 130) = 150°$

   $m\angle B = \frac{1}{2}\left(m\widehat{AD} - m\widehat{AC}\right)$

   $m\angle B = \frac{1}{2}(150 - 80)$

   $m\angle B = \frac{1}{2}(70) = 35°$

9. Let $m\overarc{AC} = m\overarc{CD} = x$. Then $m\overarc{AD} = 360 - 2x$.

$$m\angle B = \frac{1}{2}\left(m\overarc{AD} - m\overarc{AC}\right)$$
$$40 = \frac{1}{2}(360 - 2x - x)$$
$$80 = 360 - 3x$$
$$-280 = -3x$$
$$x = 93\frac{1}{3}$$
$$m\overarc{AC} = m\overarc{DC} = 93\frac{1}{3}^{\circ}$$
$$m\overarc{AD} = 173\frac{1}{3}^{\circ}$$

10. Let $m\overarc{AC} = x$; $m\overarc{AD} = 290 - x$

$$m\angle B = \frac{1}{2}\left(m\overarc{AD} - m\overarc{AC}\right)$$
$$35 = \frac{1}{2}(290 - x - x)$$
$$70 = 290 - 2x$$
$$-220 = -2x$$
$$x = 110$$
$$m\overarc{AC} = 110^{\circ} \text{ and } m\overarc{AD} = 80^{\circ}$$

11. If $m\angle 1 = 46^{\circ}$, then $m\overarc{BC} = 92^{\circ}$.

If $\overline{AC}$ is a diameter, $m\overarc{AB} = 88^{\circ}$.

$m\angle 2 = 44^{\circ}$; $m\angle 3 = 90^{\circ}$; $m\angle 4 = 46^{\circ}$;

$m\angle 5 = 44^{\circ}$.

12. If $m\angle 5 = 40^{\circ}$, then $m\overarc{AB} = 80^{\circ}$.

If $\overline{AC}$ is a diameter, $m\overarc{BC} = 100^{\circ}$.

$m\angle 1 = 50^{\circ}$; $m\angle 2 = 40^{\circ}$; $m\angle 3 = 90^{\circ}$;

$m\angle 4 = 50^{\circ}$.

13. (12, 16, 20) is a multiple of the Pythagorean Triple (3, 4, 5). Hence, half of the chord has a length of 12 and the chord has length 24.

14. The radius of the circle has length 10 using the Pythagorean Triple (6, 8, 10).

15. A

16. S

17. N

18. S

19. A

20. N

21. A

22. N

23. a. $m\angle AEB = \frac{1}{2}\left(m\overarc{AB} + m\overarc{CD}\right)$

$$75 = \frac{1}{2}\left(80 + m\overarc{CD}\right)$$
$$150 = 80 + m\overarc{CD}$$
$$m\overarc{CD} = 70^{\circ}$$

b. $m\angle BED = \frac{1}{2}\left(m\overarc{AC} + m\overarc{BD}\right)$

$$45 = \frac{1}{2}\left(62 + m\overarc{BD}\right)$$
$$90 = 62 + m\overarc{BD}$$
$$m\overarc{BD} = 28^{\circ}$$

c. $m\angle P = \frac{1}{2}\left(m\overarc{AB} - m\overarc{CD}\right)$

$$24 = \frac{1}{2}\left(88 - m\overarc{CD}\right)$$
$$48 = 88 + m\overarc{CD}$$
$$m\overarc{CD} = 40^{\circ}$$
$$m\angle CED = \frac{1}{2}\left(m\overarc{AB} + m\overarc{CD}\right)$$
$$m\angle CED = \frac{1}{2}(88 + 40) = \frac{1}{2}(128) = 64^{\circ}$$

d. $m\angle CED = \frac{1}{2}\left(m\overarc{AB} + m\overarc{CD}\right)$

$$41 = \frac{1}{2}\left(m\overarc{AB} + 20\right)$$
$$82 = m\overarc{AB} + 20$$
$$m\overarc{AB} = 62^{\circ}$$
$$m\angle P = \frac{1}{2}\left(m\overarc{AB} - m\overarc{CD}\right)$$
$$m\angle P = \frac{1}{2}(62 - 20) = \frac{1}{2}(42) = 21^{\circ}$$

e. $m\angle AEB = \frac{1}{2}\left(m\overarc{AB} + m\overarc{CD}\right)$ and

$m\angle P = \frac{1}{2}\left(m\overarc{AB} - m\overarc{CD}\right)$.

$$65 = \frac{1}{2}\left(m\overarc{AB} + m\overarc{CD}\right)$$
$$25 = \frac{1}{2}\left(m\overarc{AB} - m\overarc{CD}\right)$$

$$130 = m\overarc{AB} + m\overarc{CD}$$
$$\underline{50 = m\overarc{AB} - m\overarc{CD}}$$
$$180 = 2 \cdot m\overarc{AB}$$
$$m\overarc{AB} = 90^{\circ}; \ m\overarc{CD} = 90^{\circ}$$

23. f. $m\angle CED = \frac{1}{2}\left(m\widehat{AB} - m\widehat{CD}\right)$

$50 = \frac{1}{2}\left(m\widehat{AB} - m\widehat{CD}\right)$

$100 = \left(m\widehat{AB} - m\widehat{CD}\right)$

$m\widehat{AC} + m\widehat{BD} = 360 - 100 = 260°$

24. a. Let $BC = x$.

$6^2 = 12 \cdot x$

$36 = 12x$

$x = 3$

$BC = 3$

b. Let $DG = x$.

$4 \cdot 6 = x \cdot 3$

$24 = 3x$

$x = 8$

$DG = 8$

c. Let $CE = x$.

$3 \cdot x = 4 \cdot 12$

$3x = 48$

$x = 16$

$CE = 16$

d. Let $GE = x$.

$10 \cdot x = 5 \cdot 8$

$10x = 40$

$x = 4$

$GE = 4$

e. Let $BC = x$ and $CA = x + 5$.

$6^2 = x(x + 5)$

$36 = x^2 + 5x$

$0 = x^2 + 5x - 36$

$0 = (x + 9)(x - 4)$

$x = -9$ or $x = 4$; reject $x = -9$.

$BC = 4$.

f. Let $DG = x$ and $AG = 9 - x$.

$x(9 - x) = 4 \cdot 2$

$9x - x^2 = 8$

$0 = x^2 - 9x + 8$

$0 = (x - 8)(x - 1)$

$x = 8$ or $x = 1$;

$GD = 8$ or $GD = 1$.

g. Let $CD = ED = x$ and $CE = 2x$.

$x(2x) = 3 \cdot 30$

$2x^2 = 90$

$x^2 = 45$

$x = \sqrt{45} = 3\sqrt{5}$

$ED = 3\sqrt{5}$

h. Let $CD = x$ and $CE = x + 12$

$x(x + 12) = 5 \cdot 9$

$x^2 + 12x = 45$

$x^2 + 12x - 45 = 0$

$(x + 15)(x - 3) = 0$

$x = -15$ or $x = 3$; reject $x = -15$;

$CD = 3$.

i. Let $FC = x$

$x^2 = 4 \cdot 12$

$x^2 = 48$

$x = \sqrt{48} = 4\sqrt{3}$

$FC = 4\sqrt{3}$

j. Let $CD = x$ and $CE = x + 9$.

$x(x + 9) = 6^2$

$x^2 + 9x = 36$

$x^2 + 9x - 36 = 0$

$(x + 12)(x - 3) = 0$

$x = -12$ or $x = 3$; reject $x = -12$;

$CD = 3$

25. $5x + 4 = 2x + 19$

$3x = 15$

$x = 5$

$OE = 5(5) + 4 = 29$

26. $x(x - 2) = x + 28$

$x^2 - 2x - x - 28 = 0$

$x^2 - 3x - 28 = 0$

$(x - 7)(x + 4) = 0$

$x = 7$ or $x = -4$. If $x = 7$, then

$AC = 7 + 28 = 35$; $DE = 17\frac{1}{2}$.

If $x = -4$, then $AC = 24$; $DE = 12$.

27. Given: $\overline{DC}$ is tangent to circles B and A at points D and C, respectively.

Prove: $AC \cdot ED = CE \cdot BD$

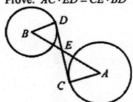

Proof: If $\overline{DC}$ is tangent to circles B and A at points D and C, then $\overline{BD} \perp \overline{DC}$ and $\overline{AC} \perp \overline{DC}$. $\angle$s D and C are congruent since they are right angles. $\angle DEB \cong \angle CEA$ because of vertical angles. $\triangle BDE \sim \triangle ACE$ by AA. It follows that $\frac{AC}{CE} = \frac{BD}{ED}$ since corresponding sides are proportional. Hence, $AC \cdot ED = CE \cdot BD$.

28. Given: $\odot O$ with $\overline{EO} \perp \overline{BC}$,
$\overline{DO} \perp \overline{BA}$, $\overline{EO} \cong \overline{OD}$

Prove: $\overparen{BC} \cong \overparen{BA}$

Proof: In $\odot O$, if $\overline{EO} \perp \overline{BC}$, $\overline{DO} \perp \overline{BA}$, and $\overline{EO} \cong \overline{OD}$, then $\overline{BC} \cong \overline{BA}$. (Chords equidistant from the center of the circle are congruent.) It follows then that $\overparen{BC} \cong \overparen{BA}$.

29. Given: $\overline{AP}$ and $\overline{BP}$ are tangent to $\odot Q$ at A and B; C is the midpoint of $\overparen{AB}$.

Prove: $\overrightarrow{PC}$ bisects $\angle APB$

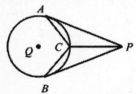

Proof: If $\overline{AP}$ and $\overline{BP}$ are tangent to $\odot Q$ at A and B, then $\overline{AP} \cong \overline{BP}$. $\overparen{AC} \cong \overparen{BC}$ since C is the midpoint of $\overparen{AB}$. It follows that $\overline{AC} \cong \overline{BC}$ and using $\overline{CP} \cong \overline{CP}$, we have $\triangle ACP \cong \triangle BCP$ by SSS. $\angle APC \cong \angle BCP$ by CPCTC and hence $\overline{PC}$ bisects $\angle APB$.

30. If $m\overparen{AD} = 136°$ and $\overparen{AC}$ is a diameter, then $m\overparen{DC} = 44°$. If $m\overparen{BC} = 50°$, then $m\overparen{AB} = 130°$.
$m\angle 1 = 93°$; $m\angle 2 = 25°$; $m\angle 3 = 43°$;
$m\angle 4 = 68°$; $m\angle 5 = 90°$; $m\angle 6 = 22°$;
$m\angle 7 = 68°$; $m\angle 8 = 22°$; $m\angle 9 = 50°$;
$m\angle 10 = 112°$

31. Each side of the square has length $6\sqrt{2}$. Therefore, the perimeter is $24\sqrt{2}$ cm.

32. The perimeter of the triangle is $15 + 5\sqrt{3}$ cm.

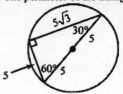

33.
$$(35 - x)^2 + (6 + x)^2 = 29^2$$
$$1225 - 70x + x^2 + 36 + 12x + x^2 = 841$$
$$2x^2 - 58x + 420 = 0$$
$$x^2 - 29x + 210 = 0$$
$$(x - 15)(x - 14) = 0$$
$x = 15$ or $x = 14$
The lengths of the segments on the hypotenuse are 14 and 15.

34. Let $AD = x = AF$; $BE = y = DB$;
$FC = z = CE$, then
$$x + y = 9$$
$$x \quad + z = 10$$
$$\quad y + z = 13$$
Subtracting the second equation from the first we get $y - z = 1$. Using this one along with the third equation, we have
$$y - z = -1$$
$$y + z = 13.$$
Adding, we get $2y = 12$ or $y = 6$.
Solving for x and z, $x = 3$ and $z = 7$.
$AD = 3$; $BE = 6$; $FC = 7$.

35. **a.** $AB > CD$

    b. $QP < QR$

    c. $m\angle A < m\angle C$

36. **a.**

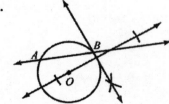

    b.

37.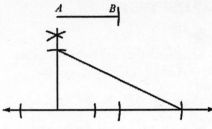

38.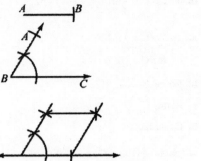

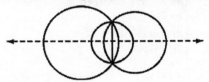

39. The locus of the midpoints of the radii of a circle is another circle.

40. The locus of the centers of all circles passing through two given points is the perpendicular bisector of the segment joining the 2 given points.

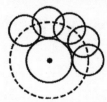

41. The locus of the centers of a penny that rolls around a half-dollar is a circle.

42. The locus of points in space less than three units from a given point is the interior of a sphere.

43. The locus of points equidistant from 2 parallel planes is a parallel plane midway between the 2 planes.

44.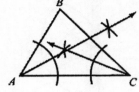

45.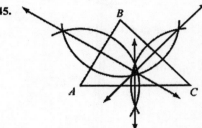

46.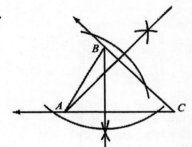

47.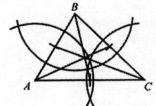

48.

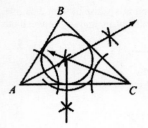

49.

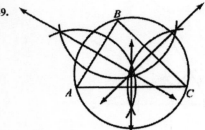

50. **a.** $BG = \frac{2}{3}(BF) = \frac{2}{3}(18) = 12$

   b. $AG = \frac{2}{3}(AE)$
   $4 = \frac{2}{3}(AE)$
   $AE = \frac{3}{2}(4) = 6$
   $GE = 2$

   c. $CG = \frac{2}{3}(DC)$
   $4\sqrt{3} = \frac{2}{3}(DC)$
   $DC = \frac{3}{2}(4\sqrt{3}) = 6\sqrt{3}$
   $DG = 2\sqrt{3}$

51. $AG = 2(GE)$ and $BG = 2(GF)$
   $2x + 2y = 2(2x - y)$ and $3y + 1 = 2(x)$.
   Simplifying: $2x + 2y = 4x - 2y$ and $3y + 1 = 2x$.
   $-2x + 4y = 0$ and $2x - 3y = 1$.
   Adding the above two equations gives $y = 1$.
   Solving for x gives
   $-2x + 4 = 0$
   $\quad -2x = -4$
   $\quad\quad x = 2$
   $BF = BG + GF = 4 + 2 = 6$;
   $AE = AG + GE = 6 + 3 = 9$.

CHAPTER TEST

1. **a.** $272°$

   b. $\mathrm{m}\widehat{ACB} = 360 - \mathrm{m}\widehat{AB} = 360 - 92 = 268°$
   $\mathrm{m}\widehat{AC} = \frac{1}{2}\mathrm{m}\widehat{ACB} = \frac{1}{2}(268) = 134°$

2. **a.** $69°$

   b. $\mathrm{m}\angle BAC = \frac{1}{2}\mathrm{m}\widehat{BC} = \frac{1}{2}(64) = 32°$

3. **a.** $\mathrm{m}\angle BAC = \frac{1}{2}\mathrm{m}\widehat{BC}$
   $\mathrm{m}\widehat{BC} = 2(24) = 48°$

   b. Isosceles

4. **a.** Right

   b. Congruent

5. **a.** $\mathrm{m}\angle 1 = \frac{1}{2}(106 + 32) = \frac{1}{2}(138) = 69°$

   b. $\mathrm{m}\angle 2 = \frac{1}{2}(106 - 32) = \frac{1}{2}(74) = 37°$

6. **a.** $214°$

   b. $\mathrm{m}\angle 3 = \frac{1}{2}(214 - 146) = \frac{1}{2}(68) = 34°$

7. **a.** Concentric

   b. $(QV)^2 = (QR)^2 + (RV)^2$
   $(RV)^2 = (QV)^2 - (QR)^2$
   $(RV)^2 = 5^2 - 3^2$
   $(RV)^2 = 25 - 9$
   $(RV)^2 = 16$
   $\quad RV = 4$
   $\quad TV = 8$

8. **a.** 1

   b. 2

9. **a.** $HP = 4$, $PJ = 5$, and $PM = 2$.
   Let $LP = x$.
   $4 \cdot 5 = 2 \cdot x$
   $\quad 20 = 2x$
   $\quad\quad x = 10$
   $\quad LP = 10$

   b. $HP = x + 1$, $PJ = x - 1$, $LP = 8$, and
   $PM = 3$.
   $(x + 1)(x - 1) = 8 \cdot 3$
   $\quad\quad x^2 - 1 = 24$
   $\quad\quad x^2 - 25 = 0$
   $(x - 5)(x + 5) = 0$
   $x = 5$ or $x = -5$; reject $x = -5$.

10.

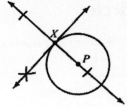

11. a. m$\angle AQB >$ m$\angle CQD$

  b. $AB > CD$

12. a. The circle with center P and radius length 3 centimeters

  b. The sphere with center P and radius length 3 centimeters

13. a. The line that is the perpendicular bisector of $\overline{AB}$

  b. The plane that is the perpendicular bisector of $\overline{AB}$

14. a. Incenter

  b. Centroid

15. a. If $MQ = 12$, then $QB = 6$.
      $MB = 12 + 6 = 18$.

  b. $\frac{1}{2}PQ = QC$
      $\frac{1}{2}(x + 4) = x$
      $x + 4 = 2x$
      $x = 4$

16. S1. In $\odot O$, chords $\overline{AD}$ and $\overline{BC}$ intersect at E

  R1. Given

  R2. Vertical angles are congruent

  R4. AA

  S5. $\dfrac{AE}{CE} = \dfrac{BE}{DE}$

Chapter 7: Areas of Polygons and Circles

SECTION 7.1: Area and Initial Postulates

1. Two triangles with equal areas are not necessarily congruent.

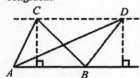

$\triangle ABC \not\cong \triangle ABD$

Two squares with equal areas must be congruent because the sides will be congruent.

5. The altitudes to $\overline{PN}$ and to $\overline{MN}$ are congruent.

This follows from the fact that $\triangle$s QMN and QPN are congruent. Corresponding altitudes of $\cong$ $\triangle$s are $\cong$.

9. $A = bh$
$A = 6 \cdot 9$
$A = 54 \text{ cm}^2$

13. $A = bh$
$A = 12 \cdot 6$
$A = 72 \text{ in}^2$

17. Using the Pythagorean Triple, (5, 12, 13), $b = 12$.

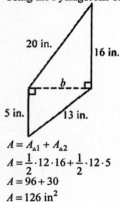

$A = A_{\triangle 1} + A_{\triangle 2}$
$A = \frac{1}{2} \cdot 12 \cdot 16 + \frac{1}{2} \cdot 12 \cdot 5$
$A = 96 + 30$
$A = 126 \text{ in}^2$

21. In the right $\triangle$, $h = 8$, using the Pythagorean Triple (6, 8, 10).

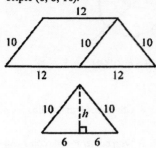

$A = A_{\text{PARALLELOGRAM}} + A_{\triangle}$
$A = 12 \cdot 8 + \frac{1}{2} \cdot 12 \cdot 8$
$A = 96 + 48$
$A = 144 \text{ units}^2$

25. $A = \frac{1}{2}bh + bh$

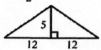

a. $A = \frac{1}{2} \cdot 24 \cdot 5 + 24 \cdot 10$
$A = 300 \text{ ft}^2$

b. 3 gallons

c. $46.50

29. a. 3 ft = 1 yd $\therefore$ 9 sq ft = 1 sq yd

b. 36 in = 1 yd $\therefore$ 1296 sq in = 1 sq yd

33. Given: $\triangle ABC$ with midpoints M, N, and P

Explain why $A_{ABC} = 4 \cdot A_{MNP}$

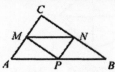

Explanation: $\overline{MN}$ joins the midpoints of $\overline{CA}$ and $\overline{CB}$ so $MN = \frac{1}{2}(AB)$. $\therefore \overline{AP} \cong \overline{PB} \cong \overline{MN}$.

$\overline{PN}$ joins the midpoints of $\overline{CB}$ and $\overline{AB}$ so $PN = \frac{1}{2}(AC)$. $\therefore \overline{AM} \cong \overline{MC} \cong \overline{PN}$.

$\overline{MP}$ joins the midpoints of $\overline{AB}$ and $\overline{AC}$ so $MP = \frac{1}{2}(BC)$. $\therefore \overline{CN} \cong \overline{NB} \cong \overline{MP}$. The four triangles are all $\cong$ by SSS. Therefore, the area of each triangle is the same. Hence the area of the big triangle is the same as four times the area of one of the smaller triangles.

37. $A = \frac{1}{2}bh$

$40 = \frac{1}{2} \cdot x(x+2)$

Multiplying by 2,

$80 = x(x+2)$

$80 = x^2 + 2x$

$0 = x^2 + 2x - 80$

$0 = (x+10)(x-8)$

$x + 10 = 0$ or $x - 8 = 0$

       $x = -10$ or $x = 8$

Reject $x = -10$.

41.
$$A_{\text{RECT.}} = bh$$
$$A_{\text{LARGE RECT.}} = (b + 0.2b)(h + 0.3h)$$
$$= (1.2b)(1.3h) = 1.56bh$$
$$A_{\text{LARGE RECT.}} - A_{\text{RECT.}} = 1.56bh - bh = 0.56bh$$

The area is increased by 56%.

45. $(a+b)(c+d) = ac + ad + bc + bd$

49. $A_{MNPQ} = (QR)(MN)$

But $MN = QP = 10$

$\therefore A_{MNPQ} = 4(10) = 40$ units^2

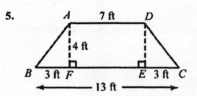

$A_{MNPQ} = (QS)(PN)$

But $PN = QM = 5$ so

$A_{MNPQ} = (QS)5$

$40 = 5(QS)$

$QS = 8$ units

SECTION 7.2: Perimeter and Area of Polygons

1. $c = 13$ using the Pythagorean Triple (5, 12, 13).

$P = a + b + c$

$P = 5 + 12 + 13$

$P = 30$ in.

5.

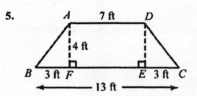

Draw the altitude from D to $\overline{BC}$. $FE = 7$ and $BF = ED = 3$. $AB = DC = 5$. Therefore, the perimeter of isosceles trapezoid $ADCB$ with altitudes $\overline{AF}$ and $\overline{DE}$

$P = 7 + 5 + 5 + 13 = 30$ ft.

9. $A = \sqrt{s(s-a)(s-b)(s-c)}$ where

$a = 13$, $b = 14$, $c = 15$ and

$s = \frac{1}{2}(13 + 14 + 15)$

$s = \frac{1}{2}(42) = 21 \therefore$

$A = \sqrt{21(21-13)(21-14)(21-15)}$

$A = \sqrt{21(8)(7)(6)}$

$A = \sqrt{(3 \cdot 7) \cdot (2 \cdot 4) \cdot 7 \cdot (2 \cdot 3)}$

$A = \sqrt{2^2 \cdot 3^2 \cdot 2^2 \cdot 7^2}$

$A = 2 \cdot 3 \cdot 2 \cdot 7$

$A = 84$ in^2

13. $A = \frac{1}{2}h(b_1 + b_2)$

$A = \frac{1}{2} \cdot 4(7 + 13)$

$A = \frac{1}{2} \cdot 4(20)$

$A = 40$ ft^2

17.

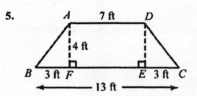

$A = \frac{1}{2} \cdot d_1 \cdot d_2$

$A = \frac{1}{2} \cdot 12(6 + 6\sqrt{3})$

$A = 6(6 + 6\sqrt{3})$

$A = (36 + 36\sqrt{3})$ units^2

21. $A = \frac{1}{2}h(b_1 + b_2)$

$$96 = \frac{1}{2} \cdot 8(b + 9)$$

$$96 = 4(b + 9)$$

$$96 = 4b + 36$$

$$60 = 4b$$

$$b = 15 \text{ cm}$$

25. Using Heron's Formula, the semiperimeter is $\frac{1}{2}(3s)$ or $\frac{3s}{2}$. Then

$$A = \sqrt{\frac{3s}{2}\left(\frac{3s}{2} - s\right)\left(\frac{3s}{2} - s\right)\left(\frac{3s}{2} - s\right)}$$

$$A = \sqrt{\frac{3s}{2}\left(\frac{s}{2}\right)\left(\frac{s}{2}\right)\left(\frac{s}{2}\right)}$$

$$A = \sqrt{\frac{3s^4}{16}} = \frac{\sqrt{3} \cdot \sqrt{s^4}}{\sqrt{16}}$$

$$A = \frac{s^2\sqrt{3}}{4}$$

29. In $\triangle ABC$, $AC = 13$ (by using the Pythagorean Theorem). Now $EC = 12$.

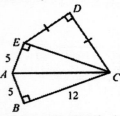

Using the 45°-45°-90° relationship,

$$DE \cdot \sqrt{2} = 12$$

$$DE = \frac{12}{\sqrt{2}} \cdot \frac{\sqrt{2}}{\sqrt{2}} = \frac{12\sqrt{2}}{2} = 6\sqrt{2}$$

$\therefore DC = 6\sqrt{2}$ (since $\overline{DC} \cong \overline{DE}$)

$$A_{ABCDE} = A_{ABC} + A_{AEC} + A_{EDC}$$

$$A_{ABCDE} = \frac{1}{2} \cdot 5 \cdot 12 + \frac{1}{2} \cdot 5 \cdot 12 + \frac{1}{2} \cdot (6\sqrt{2})(6\sqrt{2})$$

$$A_{ABCDE} = 30 + 30 + \frac{1}{2}(36)(2)$$

$$A_{ABCDE} = 30 + 30 + 36 = 96 \text{ units}^2$$

33. a. $P = 2b + 2h$

$$P = 2(245) + 2(140)$$

$$P = 490 + 280$$

$$P = 770 \text{ ft}$$

    b. Cost = ($0.59)(770)

        Cost = $454.30

37. The largest area occurs when the rectangle is a square with sides of length 10 inches.

41.

Using Pythagorean Triple (5, 12, 13), $AC = 13$.

$A_{\triangle ABC} = \sqrt{s(s - a)(s - b)(s - c)}$ where

$a = 5$, $b = 12$, $c = 13$ and

$$s = \frac{1}{2}(5 + 12 + 13)$$

$s = \frac{1}{2}(30) = 15 \ \therefore$

$$A_{\triangle ABC} = \sqrt{15(15 - 5)(15 - 12)(15 - 13)}$$

$$A_{\triangle ABC} = \sqrt{15(10)(3)(2)}$$

$$A_{\triangle ABC} = \sqrt{(3 \cdot 5) \cdot (2 \cdot 5) \cdot 3 \cdot 2}$$

$$A_{\triangle ABC} = \sqrt{2^2 \cdot 3^2 \cdot 5^2}$$

$$A_{\triangle ABC} = 2 \cdot 3 \cdot 5$$

$$A_{\triangle ABC} = 30 \text{ in}^2$$

Since $A_{\triangle ABC} \cong A_{\triangle ADC}$, then

$$A_{ABCD} = A_{\triangle ABC} + A_{\triangle ADC}$$

$$= 30 + 30 = 60 \text{ in}^2$$

45. The area of a trapezoid $= \frac{1}{2}h(b_1 + b_2)$

$$= h \cdot \frac{1}{2}(b_1 + b_2)$$

$$= h \cdot m$$

since the median, m, of a trapezoid $= \frac{1}{2}(b_1 + b_2)$.

49. $A = \frac{1}{2}d^2 = \frac{1}{2}\left(\sqrt{10}\right)^2 = \frac{1}{2}(10) = 5 \text{ in}^2$

SECTION 7.3: Regular Polygons and Area

1. First, construct the angle-bisectors of two consecutive angles, say A and B. The point of intersection, O, is the center of the inscribed circle.

Second, construct the line segment $\overline{OM}$ which is perpendicular to $\overline{AB}$. Then, using the radius $r = OM$, construct the inscribed circle with the center O.

5. In $\odot O$, draw diameter $\overline{AB}$.

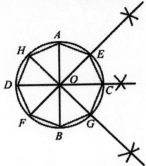

Now construct the diameter $\overline{CD}$ which is $\perp$ to $\overline{AB}$ at O. Construct the angle bisectors for $\angle$s AOC and BOC, and extend these to form diameters $\overline{EF}$ and $\overline{GH}$. Joining in the order A, E, C, G, B, F, D, and H, determines a regular octagon inscribed in $\odot O$.

9. $P = 8(3.4) = 27.2$ in.

13. In square $ABCD$, apothem $a = 5$ in. (using the $45°$-$45°$-$90°$ relationship).

17. $c = \dfrac{360}{n}$

    a. $120°$

    b. $90°$

    c. $72°$

    d. $60°$

21. $A = \dfrac{1}{2}aP = \dfrac{1}{2}(5.2)(37.5) = 97.5$ cm^2

25.

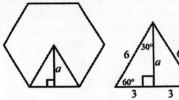

The length of the apothem is $a = 3\sqrt{3}$ (using the $30°$-$60°$-$90°$ relationship).

$A = \dfrac{1}{2}aP$ becomes

$A = \dfrac{1}{2} \cdot 3\sqrt{3} \cdot 36$

$A = 54\sqrt{3}$ cm^2

29. $P = 12(2) = 24$

$A = \dfrac{1}{2}aP$

$A = \dfrac{1}{2}(2 + \sqrt{3})24$

$A = 12(2 + \sqrt{3})$

$A = (24 + 12\sqrt{3})$ in^2

33.

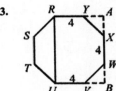

$\triangle YAX \cong \triangle WBV$

$\overline{YA} \cong \overline{AX} = \dfrac{4}{\sqrt{2}} \cdot \dfrac{\sqrt{2}}{\sqrt{2}} = \dfrac{4\sqrt{2}}{2} = 2\sqrt{2}$ (using the $45°$-$45°$-$90°$ relationship.)

$RU = 2\sqrt{2} + 4 + 2\sqrt{2}$

$\qquad = 4 + 4\sqrt{2}$

37. If a circle is divided into n congruent arcs $(n \geq 3)$, the chords determined by joining consecutive endpoints of these arcs form a regular polygon.

Proof: Let A, B, C, D, E, be the points which determine congruent arcs $\overparen{AB}$, $\overparen{BC}$, $\overparen{CD}$, $\overparen{DE}$, ... Now $\overline{AB} \cong \overline{BC} \cong \overline{CD} \cong \overline{DE} \cong \ldots$ since congruent arcs have congruent chords. Because

$m\overparen{ABC} = m\overparen{AB} + m\overparen{BC} = 2m\overparen{AB} = 2m\overparen{BC}$,

$m\overparen{BCD} = m\overparen{BC} + m\overparen{CD} = 2m\overparen{BC} = 2m\overparen{CD}$

$m\overparen{CDE} = m\overparen{CD} + m\overparen{DE} = 2m\overparen{CD} = 2m\overparen{DE}$

it follows that

$m\overparen{ABC} = m\overparen{BCD} = m\overparen{CDE} = \ldots$

Now

$m\overparen{AEC} = 360° - m\overparen{ABC}$

$m\overparen{BED} = 360° - m\overparen{BCD}$

$m\overparen{CAE} = 360° - m\overparen{CDE}$

So that $m\overparen{AEC} = m\overparen{BED} = m\overparen{CAE} = \ldots$

But $m\angle B = m\overparen{AEC}$

$\quad\quad m\angle C = m\overparen{BED}$

$\quad\quad\quad m\angle D = m\overparen{CAE} = \ldots$

so that $m\angle B = m\angle C = m\angle D = \ldots$

Thus the polygon is a regular polygon.

SECTION 7.4: Circumference and Area of a Circle

1. $C = 2\pi r \quad\quad A = \pi r^2$

$C = 2 \cdot \pi \cdot 8 \quad A = \pi \cdot 8^2$

$C = 16\pi$ cm $\quad A = 64\pi$ cm^2

5. a. $C = 2\pi r$

$44\pi = 2\pi r$

$\dfrac{44\pi}{2\pi} = r$

$r = 22$ in.

$\therefore d = 44$ in.

b. $C = 2\pi r$

$60\pi = 2\pi r$

$\dfrac{60\pi}{2\pi} = r$

$r = 30$ ft

$\therefore d = 60$ ft

9. $\ell = \dfrac{m}{360} \cdot C$

$\ell = \dfrac{60}{360} \cdot 2 \cdot \pi \cdot 8$

$\ell = \dfrac{1}{6} \cdot 16\pi$

$\ell = \dfrac{8}{3}\pi$ in.

13. $A = \pi r^2$

$143 = \pi r^2$

$\dfrac{143}{\pi} = r^2$

$\quad r \approx 6.7$ cm

17. The maximum area of a rectangle occurs when it is a square. $\therefore A = 16$ sq in.

21. $A = A_{\text{CIRCLE}} - A_{\text{SQUARE}}$

$A = \pi\left(4\sqrt{2}\right)^2 - 8^2$

$A = (32\pi - 64)$ in^2

25. $A = \pi r^2$

$154 = \pi r^2$

$r^2 = \dfrac{154}{\pi}$

$\quad r \approx 7$ cm

29. Given: Concentric circles with radii of lengths R and r, where $R > r$.

Explain: $A_{\text{RING}} = \pi(R + r)(R - r)$

$A = A_{\text{LARGER CIRCLE}} - A_{\text{SMALLER CIRCLE}}$

$A = \pi R^2 - \pi r^2$

$A = \pi\left(R^2 - r^2\right)$.

But $R^2 - r^2$ is a difference of 2 squares, so that

$A = \pi(R + r)(R - r)$.

33. a. $A = \pi r^2$

$A = \pi\left(8^2\right)$

$A \approx 201.06$ ft^2

b. $\dfrac{201.06}{70} \approx 2.87$ pints

Thus, 3 pints need to be purchased.

c. $3 \times \$2.95 = \$8.85$

37. $P_{\text{POLYGON}} \approx C_{\text{CIRCLE}}$

$P \approx 2\pi r$

$P \approx 2 \cdot \pi \cdot 7$

$P \approx 43.98$ cm

43.

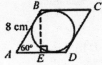

$BE = 4\sqrt{3}$ cm (using the 30°-60°-90° relationship).

$\overline{BE}$ is the same as the diameter of the circle, so then the radius of the circle = $2\sqrt{3}$ cm.

$$A_{\text{CIRCLE}} = \pi r^2 = \pi\left(2\sqrt{3}\right)^2 = \pi \cdot 4 \cdot 3 = 12\pi \text{ cm}^2$$

SECTION 7.5: More Area Relationships in the Circle

1. $P_{\text{SECTOR}} = 10 + 10 + 14 = 34$ in.

5. $A_{\vartriangle} = \dfrac{1}{2}rP$

$A_{\vartriangle} = \dfrac{1}{2}r(3s)$

$A_{\vartriangle} = \dfrac{3}{2}rs$

9. $A_{\vartriangle} = \dfrac{1}{2}rP$

$A_{\vartriangle} = \dfrac{1}{2}(2)(6+8+10)$

$A_{\vartriangle} = 24$ sq in.

13. $P = 2r + \ell$

$P = 2(8) + \dfrac{8}{3}\pi$

$P = \left(16 + \dfrac{8}{3}\pi\right)$ in.

$A = \dfrac{m}{360} \cdot \pi r^2$

$A = \dfrac{60}{360} \cdot \pi\left(8^2\right)$

$A = \dfrac{1}{6} \cdot 64\pi$

$A = \dfrac{32}{3}\pi$ in²

17. $P = AB + \ell\widehat{AB}$

$P = 12 + \dfrac{60}{360} \cdot 2\pi(12)$

$P = 12 + \dfrac{1}{6}(24\pi)$

$P = (12 + 4\pi)$ in.

$A = A_{\text{SECTOR}} - A_{\vartriangle}$

$A = \dfrac{1}{6}(\pi)(12^2) - \dfrac{12^2}{4}\sqrt{3}$

$A = \dfrac{1}{6}(144\pi) - \dfrac{144}{4}\sqrt{3}$

$A = \left(24\pi - 36\sqrt{3}\right)$ in²

21. $A = \dfrac{m}{360} \cdot \pi r^2$

$\dfrac{9}{4}\pi = \dfrac{40}{360} \cdot \pi r^2$

$\dfrac{9}{4}\pi = \dfrac{1}{9}\pi r^2$

Multiply by 9

$\dfrac{81}{4}\pi = \pi r^2$

$r^2 = \dfrac{81}{4}$

$r = \dfrac{9}{2}$ cm

25. $\ell = \dfrac{m}{360} \cdot 2\pi r$

$6\pi = \dfrac{m}{360} \cdot 2\pi(12)$

$6\pi = \dfrac{m}{360} \cdot 24\pi$

Dividing by 24π, we get

$\dfrac{1}{4} = \dfrac{m}{360}$

Multiplying by 360, we get $m = 90°$.

29. Draw in two radii to consecutive vertices of the square. Then

$r^2 + r^2 = s^2$

$2r^2 = s^2$

$r^2 = \dfrac{s^2}{2}$

$r = \dfrac{s}{\sqrt{2}} = \dfrac{s\sqrt{2}}{2}$

$A = A_{\text{CIRCLE}} - A_{\text{SQUARE}}$

$A = \pi r^2 - s^2$

$A = \pi\left(\dfrac{s\sqrt{2}}{2}\right)^2 - s^2$

$A = \pi \cdot \dfrac{s^2 \cdot 2}{4} - s^2$

$A = \left(\dfrac{\pi}{2}\right)s^2 - s^2$

33. $A_{\vartriangle} = \sqrt{s(s-a)(s-b)(s-c)}$

Also, $A_{\vartriangle} = \dfrac{1}{2}rP$ or $A_{\vartriangle} = \dfrac{1}{2}r(a+b+c)$.

So $\dfrac{1}{2}r(a+b+c) = \sqrt{s(s-a)(s-b)(s-c)}$.

Solve for r.

$r = \dfrac{2\sqrt{s(s-a)(s-b)(s-c)}}{a+b+c}$

37. $A = \dfrac{120}{360} \cdot \pi \left(18^2 - 4^2\right)$

$A = \dfrac{1}{3} \cdot \pi \cdot (324 - 16)$

$A = \dfrac{\pi}{3}(308)$

$A = \dfrac{308\pi}{3} \approx 322.54$ sq in.

CHAPTER REVIEW

1.

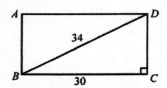

Using the Pythagorean Theorem,

$(34)^2 = (30)^2 + (DC)^2$

$1156 = 900 + (DC)^2$

$256 = (DC)^2$

$DC = 16$

$A = 30(16) = 480$ units^2

2. a. $A = 10(4) = 40$ units^2

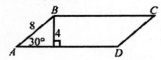

b. $A = 10\left(4\sqrt{3}\right) = 40\sqrt{3}$ units^2

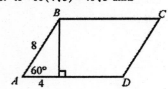

c. $A = 10\left(4\sqrt{2}\right) = 40\sqrt{2}$ units^2

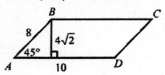

3. Using the 45°-45°-90° relationship,

$AB = DB = 5\sqrt{2}$.

$\therefore A_{ABCD} = bh$

$A_{ABCD} = \left(5\sqrt{2}\right)\left(5\sqrt{2}\right)$

$A_{ABCD} = 50$ units^2

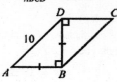

4. $A = \sqrt{s(s-a)(s-b)(s-c)}$ where

$s = \dfrac{1}{2}(17 + 25 + 26)$

$s = \dfrac{1}{2}(68)$

$s = 34$

$A = \sqrt{34(34 - 26)(34 - 25)(34 - 17)}$

$A = \sqrt{34(8)(9)(17)}$

$A = \sqrt{2^4 \cdot 3^2 \cdot 17^2}$

$A = 2^2 \cdot 3 \cdot 17$

$A = 204$ units^2

5. $A = \sqrt{s(s-a)(s-b)(s-c)}$ where

$s = \dfrac{1}{2}(26 + 28 + 30)$

$s = \dfrac{1}{2}(84)$

$s = 42$

$A = \sqrt{42(42 - 26)(42 - 28)(42 - 30)}$

$A = \sqrt{42(16)(14)(12)}$

$A = \sqrt{2^8 \cdot 3^2 \cdot 7^2}$

$A = 2^4 \cdot 3 \cdot 7$

$A = 336$ units^2

6. Using the (3, 4, 5) Pythagorean Triple,
$BH = 4$

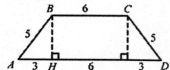

$A = \dfrac{1}{2}h\left(b_1 + b_2\right)$

$A = \dfrac{1}{2} \cdot 4(6 + 12)$

$A = 36$ units^2

7.

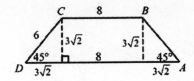

a. $A = \frac{1}{2}h(b_1 + b_2)$

$A = \frac{1}{2} \cdot 3\sqrt{2}\left(8 + (8 + 6\sqrt{2})\right)$

$A = \frac{1}{2} \cdot 3\sqrt{2}\left(16 + 6\sqrt{2}\right)$

$A = 3\sqrt{2}\left(8 + 3\sqrt{2}\right)$

$A = \left(24\sqrt{2} + 18\right)$ units^2

b. $A = \frac{1}{2} \cdot 3\left(8 + (8 + 6\sqrt{3})\right)$

$A = \frac{3}{2}\left(16 + 6\sqrt{3}\right)$

$A = \left(24 + 9\sqrt{3}\right)$ units^2

c. $A = \frac{1}{2} \cdot 3\sqrt{3}\,(8 + 14)$

$A = \frac{1}{2}(3\sqrt{3})(22)$

$A = 33\sqrt{3}$ units^2

8. $A = \frac{1}{2}d_1 \cdot d_2$

$A = \frac{1}{2}(18)(24)$

$A = 216$ in^2

Using the (9, 12, 15) Pythagorean Triple,
$P = 4(15) = 60$ in.

9. **a.** $A = (140)(160) - \left[(80)(35) + (30)(20)\right]$

$A = 22,400 - \left[2800 + 600\right]$

$A = 22,400 - 3400$

$A = 19,000$ ft^2

b. $\frac{19,000}{5000} = 3.8$ bags

Tom needs to buy 4 bags.

c. Cost $= 4 \cdot \$18 = \$72$

10. **a.** $A = (9 \cdot 8) + (12 \cdot 8)$

$A = 72 + 96$

$A = 168$ ft^2

$\frac{168}{60} = 2.8$ double rolls

3 double rolls are needed.

b. $P = 2a + 2b$

$P = 2(9) + 2(12)$

$P = 18 + 24$

$P = 42$ ft or 14 yd

$\frac{14}{5} = 2.8$ rolls

3 rolls of border are needed.

11.

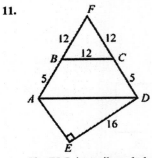

If $\triangle FBC$ is equilateral, then so is $\triangle FAD$.

$\therefore \quad AD = 17$

$(AD)^2 = (AE)^2 + (ED)^2$

$(17)^2 = (AE)^2 + (16)^2$

$28 = (AE)^2 + 256$

$(AE)^2 = 33$

$AE = \sqrt{33}$

a. $A_{EAFD} = A_{FAD} + A_{AED}$

$A_{EAFD} = \frac{17^2}{4}\sqrt{3} + \frac{1}{2}(16)\sqrt{33}$

$A_{EAFD} = \left(\frac{289}{4}\sqrt{3} + 8\sqrt{33}\right)$ units^2

b. $P_{EAFD} = AF + FD + DE + AE$

$P_{EAFD} = 17 + 17 + 16 + \sqrt{33}$

$P_{EAFD} = \left(50 + \sqrt{33}\right)$ units

12.

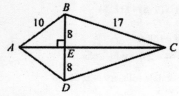

$AE = 6$
(using the 6-8-10 Pythagorean Triple)
$EC = 15$
(using the 8-15-17 Pythagorean Triple)
$A = \frac{1}{2} d_1 \cdot d_2$
$A = \frac{1}{2}(BD)(AC)$
$A = \frac{1}{2}(16)(6+15)$
$A = 8(21)$
$A = 168 \text{ units}^2$

13.

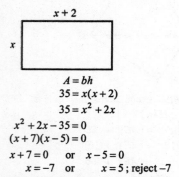

$A = bh$
$35 = x(x+2)$
$35 = x^2 + 2x$
$x^2 + 2x - 35 = 0$
$(x+7)(x-5) = 0$
$x+7 = 0 \quad \text{or} \quad x-5 = 0$
$x = -7 \quad \text{or} \qquad x = 5 \text{ ; reject } -7$
The rectangle is 5 cm by 7 cm.

14. $P = a + b + c$

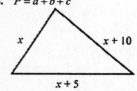

a. $60 = x + (x+10) + (x+5)$
$60 = 3x + 15$
$3x = 45$
$x = 15$
$x + 10 = 25$
$x + 5 = 20$
The three sides have lengths 15 cm, 25 cm, and 20 cm.

b. (15, 20, 25) is a Pythagorean Triple in which 25 is the length of the hypotenuse.
$A = \frac{1}{2} bh$
$A = \frac{1}{2}(20)(15)$
$A = 150 \text{ cm}^2$

15.

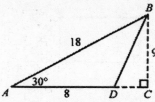

Using the 30°-60°-90° relationship, $BC = 9$.
$A = \frac{1}{2} bh$
$A = \frac{1}{2}(8)(9)$
$A = 36 \text{ units}^2$

16. $A = \frac{s^2}{4}\sqrt{3}$
$A = \frac{12^2}{4}\sqrt{3}$
$A = \frac{144}{4}\sqrt{3}$
$A = 36\sqrt{3} \text{ cm}^2$

17.

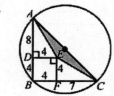

$A = A_{ABC} - \left(A_{ADE} + A_{BDEF} + A_{EFC} \right)$
$A = \frac{1}{2}(11)(12) - \left[\frac{1}{2}(8)(4) + 4^2 + \frac{1}{2}(4)(7) \right]$
$A = 66 - [16 + 16 + 14]$
$A = 66 - 46$
$A = 20 \text{ units}^2$

18. a. $c = \frac{360}{5} = 72°$

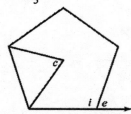

b. $i = \frac{(5-2)180}{2} = \frac{3(180)}{5} = \frac{540}{5} = 180°$

c. $e = 180° - 108° = 72°$

19.

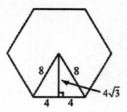

$$A = \frac{1}{2}aP$$
$$A = \frac{1}{2}(4\sqrt{3})(6 \cdot 8)$$
$$A = 96\sqrt{3} \text{ ft}^2$$

20. $P = 3(12\sqrt{3}) = 36\sqrt{3}$

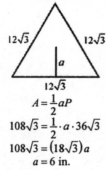

$$A = \frac{1}{2}aP$$
$$108\sqrt{3} = \frac{1}{2} \cdot a \cdot 36\sqrt{3}$$
$$108\sqrt{3} = (18\sqrt{3})a$$
$$a = 6 \text{ in.}$$

21. $x \cdot \sqrt{3} = 9$
$$x = \frac{9}{\sqrt{3}}$$
$$x = \frac{9}{\sqrt{3}} \cdot \frac{\sqrt{3}}{\sqrt{3}} = \frac{9\sqrt{3}}{3} = 3\sqrt{3}$$

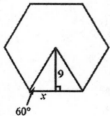

Each side has length $6\sqrt{3}$ in.
$$P = n(s)$$
$$P = 6(6\sqrt{3})$$
$$P = 36\sqrt{3} \text{ in.}$$
$$A = \frac{1}{2}aP$$
$$A = \frac{1}{2}(9)(36\sqrt{3})$$
$$A = 162\sqrt{3} \text{ in}^2$$

22. a. $c = \frac{360}{n}$
$$45 = \frac{360}{n}$$
$$45n = 360$$
$$n = 8$$

b. $P = n(s)$
$$P = 8(5)$$
$$P = 40$$
$$A = \frac{1}{2}aP$$
$$A \approx \frac{1}{2}(6)(40)$$
$$A \approx 120 \text{ cm}^2$$

23. a. No. $\perp$ bisectors of sides of a parallelogram are not necessarily concurrent.

b. No. $\perp$ bisectors of sides of a rhombus are not concurrent.

c. Yes. $\perp$ bisectors of sides of a rectangle are concurrent.

d. Yes. $\perp$ bisectors of sides of a square are concurrent.

24. a. No. $\angle$ bisectors of a parallelogram are not necessarily concurrent.

b. Yes. $\angle$ bisectors of a rhombus are concurrent.

c. No. $\angle$ bisectors of a rectangle are not necessarily concurrent.

d. Yes. $\angle$ bisectors of a square are concurrent.

25. If the radius of the inscribed circle is 7 in., then the length of each side of the triangle is $14\sqrt{3}$ in.

$$A_\triangle = \frac{1}{2}rP$$
$$A_\triangle = \frac{1}{2}(7)(3 \cdot 14\sqrt{3})$$
$$A_\triangle = 147\sqrt{3}$$
$$A_\triangle \approx 254.61 \text{ sq in.}$$

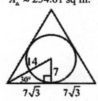

26. a. $A = (20)(30) - (12)(24)$
$$A = 600 - 288$$
$$A = 312 \text{ ft}^2$$

b. $\frac{312}{9} = 34\frac{2}{3} \text{ yd}^2$

35 yd^2 should be purchased.

c. $35 \cdot \$9.97 = \$348.95$

27. $A = A_{SQUARE} - 4 \cdot A_{SECTOR}$

$A = 8^2 - 4 \cdot \left[\dfrac{90}{360} (\pi \cdot 4^2) \right]$

$A = 64 - 4 \left[\dfrac{1}{4} (16\pi) \right]$

$A = (64 - 16\pi) \text{ units}^2$

28. $A = A_{SEMICIRCLE} - A_{TRIANGLE}$

$A = \dfrac{1}{2} (\pi \cdot 7^2) - \dfrac{1}{2} (7)(7\sqrt{3})$

$A = \left(\dfrac{49}{2} \pi - \dfrac{49}{2} \sqrt{3} \right) \text{ units}^2$

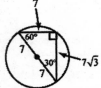

29. $A_{SEGMENT} = A_{SECTOR} - A_{TRIANGLE}$

$A_{SEGMENT} = \dfrac{60}{360} \cdot \pi \cdot 4^2 - \dfrac{4^2}{4} \sqrt{3}$

$A_{SEGMENT} = \left(\dfrac{8}{3} \pi - 4\sqrt{3} \right) \text{ units}^2$

30.

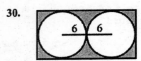

$A = A_{RECT} - 2 \cdot A_{CIRCLE}$

$A = (12)(24) - 2 \cdot (\pi \cdot 6^2)$

$A = (288 - 72\pi) \text{ units}^2$

31.

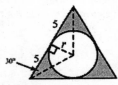

The radius of the circle is r.

$r\sqrt{3} = 5$

$r = \dfrac{5}{\sqrt{3}} = \dfrac{5\sqrt{3}}{3}$

$A = A_{EQ.\triangle} - A_{CIRCLE}$

$A = \dfrac{10^2}{4} \sqrt{3} - \pi \left(\dfrac{5\sqrt{3}}{3} \right)^2$

$A = 25\sqrt{3} - \pi \cdot \dfrac{75}{9}$

$A = \left(25\sqrt{3} - \dfrac{25}{3} \pi \right) \text{ units}^2$

32.

$\ell = \dfrac{m}{360} (2\pi r)$

$\ell = \dfrac{40}{360} (2 \cdot \pi \cdot 3\sqrt{5})$

$\ell = \dfrac{1}{9} (6\pi\sqrt{5})$

$\ell = \dfrac{2\pi\sqrt{5}}{3} \text{ cm}$

$A = \dfrac{m}{360} (\pi r^2)$

$A = \dfrac{40}{360} \left(\pi \cdot (3\sqrt{5})^2 \right)$

$A = \dfrac{1}{9} (\pi \cdot 45)$

$A = 5\pi \text{ cm}^2$

33. a. $C = \pi d$

$66 = \dfrac{22}{7} \cdot d$

$\dfrac{7}{22} \cdot 66 = \dfrac{7}{22} \cdot \dfrac{22}{7} d$

$d = 21 \text{ ft}$

b. $d = 21$, $r = \dfrac{21}{2}$

$A = \pi r^2$

$A = \dfrac{22}{7} \left(\dfrac{21}{2} \right)^2$

$A = \dfrac{693}{2} \approx 346\dfrac{1}{2} \text{ ft}^2$

34. a. $A_{SECTOR} = \dfrac{m}{360} (\pi r^2)$

$A = \dfrac{80}{360} (27\pi)$

$A = \dfrac{2}{9} (27\pi)$

$A = 6\pi \text{ ft}^2$

b. Since $\pi r^2 = 27\pi$,

$$r^2 = 27$$
$$r = \sqrt{27}$$
$$r = 3\sqrt{3}$$

$$\ell = \frac{m}{360}(2\pi r)$$
$$\ell = \frac{80}{360}(2\pi \cdot 3\sqrt{3})$$
$$\ell = \frac{2}{9}(6\pi\sqrt{3})$$
$$\ell = \frac{4\pi}{3}\sqrt{3}$$
$$P = 2r + \ell$$
$$P = 2(3\sqrt{3}) + \frac{4\pi}{3}\sqrt{3}$$
$$P = \left(6\sqrt{3} + \frac{4\pi}{3}\sqrt{3}\right) \text{ ft}$$

35.

$$A_{\text{SHADED}} = A_{\text{SEMICIRCLE}} - A_{\triangle}$$
$$A_{\text{SHADED}} = \frac{1}{2} \cdot (\pi \cdot 6^2) - \frac{1}{2} \cdot 6 \cdot 12$$
$$A_{\text{SHADED}} = 18\pi - 36$$

The area sought is one-half the shaded area.

$$A = \frac{1}{2}(18\pi - 36)$$
$$A = (9\pi - 18) \text{ in}^2$$

36. Given: Concentric circles with radii of lengths R and r with $R > r$; O is the center of the circles.

Prove: $A_{\text{RING}} = \pi(BC)^2$

Proof: By an earlier theorem,

$$A_{\text{RING}} = \pi R^2 - \pi r^2$$
$$= \pi(OC)^2 - \pi(OB)^2$$
$$= \pi\left[(OC)^2 - (OB)^2\right]$$

In rt. $\triangle OBC$,

$$(OB)^2 + (BC)^2 = (OC)^2$$
$$\therefore (OC)^2 - (OB)^2 = (BC)^2$$

In turn, $A_{\text{RING}} = \pi(BC)^2$

37. The area of a circle circumscribed about a square is twice the area of the circle inscribed within the square.

Proof: Let r represent the length of radius of the inscribed circle.
Using the 45°-45°-90° relationship, the radius of the larger circle is $r\sqrt{2}$.
Now,

$$A_{\text{INSCRIBED CIRCLE}} = \pi r^2$$
$$A_{\text{CIRCUMSCRIBED CIRCLE}} = \pi\left(r\sqrt{2}\right)^2$$
$$= \pi \cdot \left(r^2 \cdot 2\right)$$
$$= 2\left(\pi r^2\right)$$
$$= 2\left(A_{\text{INSCRIBED CIRCLE}}\right)$$

38. If semicircles are constructed on each of the sides of a right triangle, then the area of the semicircle on the hypotenuse is equal to the sum of the areas of the semicircles on the two legs.

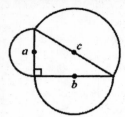

Proof: The radii of the semicircles are $\frac{1}{2}a$, $\frac{1}{2}b$,

and $\frac{1}{2}c$. The area of the semicircle on the hypotenuse is

$$A = \frac{1}{2}\left[\pi\left(\frac{1}{2}c\right)^2\right]$$

$$= \frac{1}{2} \cdot \pi\left(\frac{1}{4}c^2\right)$$

$$= \frac{1}{2}\pi\left[\frac{1}{4}\left(a^2+b^2\right)\right] \quad \left(c^2 = a^2 + b^2 \text{ by the}\right.$$

$$\text{Pythagorean Theorem.)}$$

$$= \frac{1}{2}\pi\left[\frac{1}{4}a^2 + \frac{1}{4}b^2\right]$$

$$= \frac{1}{2}\pi\left(\frac{1}{4}a^2\right) + \frac{1}{2}\pi\left(\frac{1}{4}b^2\right)$$

$$= \frac{1}{2}\pi\left(\frac{1}{2}a\right)^2 + \frac{1}{2}\pi\left(\frac{1}{2}b\right)^2$$

$$= \frac{1}{2}\left[\pi\left(\frac{1}{2}a\right)^2\right] + \frac{1}{2}\left[\pi\left(\frac{1}{2}b\right)^2\right]$$

which is the sum of areas of the semicircles on the two legs.

39. a. $A = (18)(15) - \left[\frac{1}{2}(3.14) \cdot 3^2 + \frac{1}{4}(3.14) \cdot 3^2\right]$

$A = 270 - 21.2$

$A = 248.8 \text{ ft}^2$

The approx. number of yd^2 of carpeting

needed is $\frac{248.8}{9} \approx 28$.

b. From (a), the number of ft^2 to be tiled is

21.2 ft^2.

40. a. $A = \frac{1}{2}\left[\pi R^2 - \pi r^2\right]$

$A = \frac{1}{2}\left[(3.14)(30)^2 - (3.14)(18)^2\right]$

$A = \frac{1}{2}[1808.64]$

$A = 904.32 \text{ ft}^2 \approx 905 \text{ ft}^2$

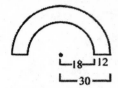

b. Cost = (905)($0.18)

Cost = $162.90

c. Length $= \frac{1}{2}(2\pi R) + \frac{1}{2}(2\pi r)$

$= \frac{1}{2}[2(3.14)(31)] + \frac{1}{2}[2(3.14)(17)]$

$= 97.34 + 52.38$

$= 150.72$

CHAPTER TEST

1. a. square inches

b. equal

2. a. $A = s^2$

b. $C = 2\pi r$

3. a. True

b. False

4. 23 cm^2

5.

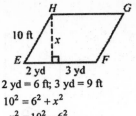

2 yd = 6 ft; 3 yd = 9 ft

$10^2 = 6^2 + x^2$

$x^2 = 10^2 - 6^2$

$x = \sqrt{100 - 36}$

$x = \sqrt{64} = 8$

$A_{EFGH} = bh = (6 + 9)(8) = 120 \text{ ft}^2$

6. $A_{MNPQ} = \frac{1}{2}d_1 d_2 = \frac{1}{2}(8)(6) = 24 \text{ ft}^2$

7. $s = \frac{1}{2}(4 + 13 + 15) = 16$

$A = \sqrt{s(s-a)(s-b)(s-c)}$

$A = \sqrt{16(16-4)(16-13)(16-15)}$

$A = \sqrt{16(12)(3)(1)}$

$A = \sqrt{4^2 \cdot 3^2 \cdot 2^2}$

$A = 4 \cdot 3 \cdot 2$

$A = 24 \text{ cm}^2$

8.

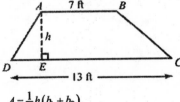

$A = \frac{1}{2}h(b_1 + b_2)$

$60 = \frac{1}{2}h(7 + 13)$

$120 = 20h$

$h = 6$

$AE = 6 \text{ ft}$

9. a. $P = ns = 5(5.8) = 29 \text{ in.}$

b. $A = \frac{1}{2}aP = \frac{1}{2}(4.0)(29) = 58 \text{ in}^2$

10. a. $C = 2\pi r = 2\pi(5) = 10\pi \text{ in.}$

b. $A = \pi r^2 = \pi(5)^2 = 25\pi \text{ in}^2$

11. $\ell\widehat{AC} = \frac{m\widehat{AC}}{360} \cdot 2\pi r$

$\ell\widehat{AC} = \frac{45}{360} \cdot 2\left(\frac{22}{7}\right)(7)$

$\ell\widehat{AC} = \frac{1}{8} \cdot 44$

$\ell\widehat{AC} = \frac{11}{2} = 5\frac{1}{2} \text{ in.}$

12. $d = 2r$

$20 = 2r$

$r = 10 \text{ cm}$

$A = \pi r^2$

$A = (3.14)(10)^2 = 314 \text{ cm}^2$

13. By the 45°-45°-90° relationship, the radius of the circle is 4.

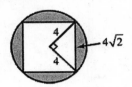

$A_{\text{SHADED}} = A_{\text{CIRCLE}} - A_{\text{SQUARE}}$

$A_{\text{SHADED}} = (\pi \cdot r^2) - s^2$

$A_{\text{SHADED}} = (\pi \cdot 4^2) - (4\sqrt{2})^2$

$A_{\text{SHADED}} = (16\pi - 32) \text{ in}^2$

14. $A = \frac{m}{360}\pi r^2$

$A = \frac{135}{360} \cdot \pi \cdot 12^2$

$A = \frac{3}{8} \cdot \pi \cdot 144$

$A = 54\pi \text{ cm}^2$

15.

$A_{\text{SHADED}} = A_{\text{SECTOR}} - A_{\text{TRIANGLE}}$

$A_{\text{SHADED}} = \frac{m}{360}\pi r^2 - \frac{1}{2}bh$

$A_{\text{SHADED}} = \frac{90}{360} \cdot \pi \cdot 12^2 - \frac{1}{2}(12)(12)$

$A_{\text{SHADED}} = \frac{1}{4} \cdot \pi \cdot 144 - 72$

$A_{\text{SHADED}} = (36\pi - 72) \text{ in}^2$

16. $P = a + b + c$

$P = 5 + 12 + 13$

$P = 30$

$A = \frac{1}{2}rP$

$30 = \frac{1}{2}r(30)$

$1 = \frac{1}{2}r$

$r = 2 \text{ in.}$

Chapter 8: Surfaces and Solids

SECTION 8.1: Prisms, Area, and Volume

1. **a.** Yes

    b. Oblique

    c. Hexagon

    d. Oblique hexagonal prism

    e. Parallelogram

5. **a.** cm^2

    b. cm^3

9. $V = bh$
    $V = 12(10)$
    $V = 120\ cm^3$

13. **a.** $2n$

    b. n

    c. $2n$

    d. $3n$

    e. n

    f. 2

    g. $n + 2$

17. **a.** $L = hP$
    $L = (6)(3 + 4 + 5)$
    $L = 72\ ft^2$

    b. $T = L + 2B$
    $T = 72 + 2\left(\frac{1}{2}bh\right)$
    $T = 72 + 4 \cdot 3$
    $T = 84\ ft^2$

    c. $V = Bh$
    $V = \left(\frac{1}{2}bh\right) \cdot 6$
    $V = \left(\frac{1}{2} \cdot 4 \cdot 3\right)(6)$
    $V = 36\ ft^3$

21.

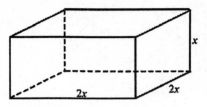

$V = l \cdot w \cdot h$
$108 = 2x \cdot 2x \cdot x$
$108 = 4x^3$
$x^3 = 27$
$x = 3 \to 2x = 6$

25. 1 foot equals 12 inches.
    $L = 5 \cdot 6 + 12 \cdot 6 + 5 \cdot 6 + 12 \cdot 6$
    $L = 30 + 72 + 30 + 7$
    $L = 204\ in^2$
    $B = 5 \cdot 12 = 60\ in^2$
    The area of both bases is $120\ in^2$.
    Cost = $0.01(204) + $0.02(120)$
    Cost = $2.04 + 2.40
    Cost = 4.44

29. **a.** $T = L + 2B$
    $T = hP + 2(e \cdot e)$
    $T = e(4e) + 2e^2$
    $T = 4e^2 + 2e^2$
    $T = 6e^2$

    b. $T = 6e^2$
    $T = 6 \cdot 4^2$
    $T = 96\ cm^2$

    c. $V = Bh$
    $V = e^2 \cdot e$
    $V = e^3$

    d. $V = e^3$
    $V = 4^3 = 64\ cm^3$

33. $V = 15' \cdot 12' \cdot 2'$
    $V = 360\ ft^3$
    $1\ yd^3 = 27\ ft^3$
    $V = 360\ ft^3 \div \dfrac{27\ ft^3}{1\ yd^3}$
    $V = 13\frac{1}{3}\ yd^3$
    $\text{Cost} = 13\frac{1}{3}\ yd^3 \cdot \dfrac{\$9.60}{1\ yd^3} = \$128$

37. $V = 2 \cdot 1 \cdot \dfrac{2}{3} = \dfrac{4}{3}$ ft^3

# of gallons $= \dfrac{4}{3} \cdot 7.5 = 10$ gal

41. $V = Bh$

$V = \left(\dfrac{1}{2}aP\right) \cdot 12$

$V = \dfrac{1}{2} \cdot 8.2 \cdot (12 \cdot 5) \cdot 12$

$V = 2952$ cm^3

SECTION 8.2: Pyramids, Area, and Volume

1. a. Right pentagonal prism

   b. Oblique pentagonal prism

5. a. Pyramid

   b. E

   c. $\overline{EA}$, $\overline{EB}$, $\overline{EC}$, $\overline{ED}$

   d. $\triangle EAB$, $\triangle EBC$, $\triangle ECD$, $\triangle EAD$

   e. No

9. $T = 12 + 16 + 12 + 10 + 16$

$T = 66$ in^2

13. a. $n + 1$

   b. n

   c. n

   d. $2n$

   e. n

   f. $n + 1$

17. a. Slant height

   b. Lateral edge

21. a. $B = \dfrac{1}{2}aP$

$B = \dfrac{1}{2}(6.3)(5 \cdot 9.2)$

$B = 144.9$ cm^2

   b. $V = \dfrac{1}{3}Bh$

$V = \dfrac{1}{3}(144.9)(14.6)$

$V = 705.18$ cm^3

25. $V = \dfrac{1}{3}Bh$

$72 = \dfrac{1}{3} \cdot x^2 \cdot x$

$72 = \dfrac{1}{3}x^3$

$x^3 = 216$

$x = 6$

$l^2 = a^2 + h^2$

$l^2 = 3^2 + 6^2$

$l^2 = 45$

$l = \sqrt{45} = 3\sqrt{5}$

$L = \dfrac{1}{2}lP$

$L = \dfrac{1}{2}(3\sqrt{5})(4 \cdot 6)$

$L = 36\sqrt{5}$

$T = L + B$

$T = 36\sqrt{5} + (6 \cdot 6)$

$T = 36\sqrt{5} + 36 \approx 116.5$ in^2

29. $V = \dfrac{1}{3}Bh$

$V = \dfrac{1}{3}\left(\dfrac{1}{2}aP\right)(15)$

$V = \dfrac{1}{3} \cdot \dfrac{1}{2} \cdot (7.5)(12 \cdot 4)(15)$

$V = 900$ ft^3

33. $V = V_{\text{TALLER PYRAMID}} - V_{\text{SHORTER PYRAMID}}$

$V = \dfrac{1}{3}B_1h_1 - \dfrac{1}{3}B_2h_2$

$V = \dfrac{1}{3}(6 \cdot 6)(32) - \dfrac{1}{3}(3 \cdot 3)(16)$

$V = 384 - 48$

$V = 336$ in^3

37.

$l^2 + 9 = 34$

$l^2 = 25$

$l = 5$

$T = B + L$

$T = 36 + \dfrac{1}{2} \cdot 24 \cdot 5$

$T = 36 + 60$

$T = 96$ in^2

41.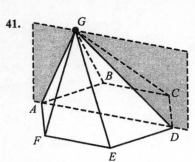

$$V_{ABCD} = V_{DEFA}$$
$$V_{ABCDEF} = V_{ABCD} + V_{DEFA}$$
$$V_{ABCDEF} = 19.7 + 19.7 = 39.4 \text{ in}^3$$

SECTION 8.3: Cylinders and Cones

1. a. Yes

   b. Yes

   c. Yes

5. $L = 2\pi rh$
$L = 2\pi(1.5)(4.25)$
$L = 12.75\pi \text{ in}^2$

$B = \pi r^2$
$B = \pi(1.5)^2$
$B = 2.25\pi$

$T = L + 2B$
$T = 12.75\pi + 2(2.25\pi)$
$T = 17.25\pi \approx 54.19 \text{ in}^3$

9.
$$L = 2\pi rh$$
$$12\pi = 2\pi r(4+1)$$
$$12\pi = 2\pi(r^2 + r)$$
$$\frac{12\pi}{2\pi} = \frac{2\pi(r^2 + r)}{2\pi}$$
$$r^2 + r = 6$$
$$r^2 + r - 6 = 0$$
$$(r+3)(r-2) = 0$$
$$r + 3 = 0 \quad \text{or} \quad r - 2 = 0$$
$$r = -3 \quad \text{or} \quad r = 2; \text{ reject } -3$$

The radius has a length of 2 inches and the altitude has a length of 3 inches.

13. $r^2 + h^2 = l^2$
$4^2 + 6^2 = l^2$
$l^2 = 52$
$l = \sqrt{52} = 2\sqrt{13} \approx 7.21 \text{ cm}$

17. $r^2 + h^2 = l^2$
$6^2 + h^2 = (2h)^2$
$3h^2 = 36$
$h^2 = 12$
$h = 2\sqrt{3}$
$l = 2h = 2(2\sqrt{3}) = 4\sqrt{3} \approx 6.93 \text{ in.}$

21. $l^2 = r^2 + h^2$
$l^2 = 7^2 + 6^2$
$l^2 = 85$
$l = \sqrt{85}$

a. $L = \frac{1}{2} l C$
$L = \frac{1}{2}\sqrt{85} \cdot 2\pi r$
$L = \frac{1}{2}\sqrt{85} \cdot 2 \cdot \pi \cdot 6$
$L = 6\pi\sqrt{85} \approx 173.78 \text{ in}^2$

b. $T = L + B$
$T = 6\pi\sqrt{85} + \pi r^2$
$T = 6\pi\sqrt{85} + \pi \cdot 6^2$
$T = 6\pi\sqrt{85} + 36\pi \approx 286.88 \text{ in}^2$

c. $V = \frac{1}{3} Bh$
$V = \frac{1}{3}\pi r^2 \cdot 7$
$V = \frac{1}{3} \cdot \pi \cdot 6^2 \cdot 7$
$V = 84\pi \approx 263.89 \text{ in}^3$

25. The solid formed is a cone with $r = 20$ and $h = 15$.
$V = \frac{1}{3} Bh$
$V = \frac{1}{3}\pi r^2 h$
$V = \frac{1}{3} \cdot \pi \cdot 20^2 \cdot 15$
$V = 2000\pi$

29. $d = 10 \therefore r = 5$

$$V = \frac{1}{3}Bh$$

$$V = \frac{1}{3}\pi r^2 h$$

$$100\pi = \frac{1}{3} \cdot \pi \cdot 5^2 \cdot h$$

$$100\pi = \frac{25\pi}{3} \cdot h$$

$$h = 12$$

$$L = \frac{1}{2}lC$$

$$L = \frac{1}{2} \cdot 13 \cdot 2\pi r$$

$$L = \frac{1}{2} \cdot 13 \cdot 2\pi \cdot 5$$

$$L = 65\pi \approx 204.2 \text{ cm}^2$$

33. $V = V_{\text{LARGE CYLINDER}} - V_{\text{SMALL CYLINDER}}$

$$V = \pi R^2 h - \pi r^2 h$$

$$V = \pi h(R^2 - r^2)$$

$$V = \pi h(R - r)(R + r)$$

37. Lateral area of smaller $= 2\pi rh$

Lateral area of larger $= 2\pi(2r)(2h) = 8\pi rh$

$$\frac{\text{Lateral area of larger}}{\text{Lateral area of smaller}} = \frac{8\pi rh}{2\pi rh} = \frac{4}{1} \text{ or } 4:1$$

41. $V = V_{\text{LARGER CONE}} - V_{\text{SMALLER CONE}}$

$$V = \frac{1}{3}\pi R^2 h - \frac{1}{3}\pi r^2 h$$

45. $V = \pi r^2 h$

$$V = \pi \cdot (1.5)^2 \cdot 6$$

$$V = 13.5\pi$$

$$V \approx 42.4115 \text{ ft}^3$$

# of gallons $= 42.4115 \cdot 7.5 \approx 318 \text{ gal}$

SECTION 8.4: Polyhedrons and Spheres

1. Polyhedron *EFGHIJK* is concave.

5. A regular hexahedron has 6 faces (*F*), 8 vertices (*V*), and 12 edges (*E*).

$$V + F = E + 2$$

$$8 + 6 = 12 + 2$$

9. Nine faces

13.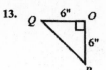

a. Using the 45°-45°-90° relationship,

$QR = 6\sqrt{2} \approx 8.49$ in.

b.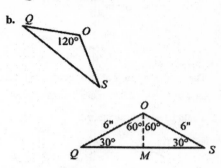

Where *M* is the midpoint of $\overline{QS}$, $OM = 3$ and $QM = 3\sqrt{3}$ by the 30°-60°-90° relationship.

$QS = 2(QM) = 2(3\sqrt{3}) = 6\sqrt{3}$.

That is, $QS = 6\sqrt{3} \approx 10.39$ in.

17. $S = 6(4.2)^2 = 105.84 \text{ cm}^2$

21. a. $A = 34.9(12) + 52.5(20) = 1468.8 \text{ cm}^2$

b. Cost $= 330(\$0.008) = \$2.64$

25.

$h = d$ or $h = 2r$

In the triangle shown, $\frac{h}{2} = r$.

Then,

$$r^2 + r^2 = 6^2$$

$$2r^2 = 36$$

$$r^2 = 18$$

$$r = \sqrt{18} = 3\sqrt{2} \approx 4.24 \text{ in.}$$

$$h = 6\sqrt{2} \approx 8.49 \text{ in.}$$

29. a. $S = 4\pi r^2$
$S = 4\pi \cdot 3^2$
$S = 36\pi \approx 113.1 \text{ m}^2$

b. $V = \frac{4}{3}\pi r^3$
$V = \frac{4}{3}\pi \cdot 3^3$
$V = 36\pi \approx 113.1 \text{ m}^3$

33. $S = 4\pi r^2$
$S = 4\pi \cdot 3^2$
$S = 36\pi \approx 113.1 \text{ ft}^2$
The number of pints of paint needed to paint the
tank is $\frac{113.1}{40} = 2.836$ pints. 3 pints of paint
would have to be purchased.

37. a. Yes

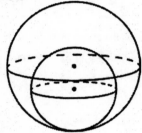

b. Yes

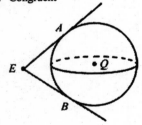

41. Congruent

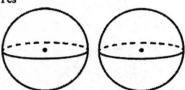

45.

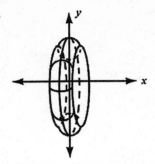

The solid of revolution looks like an inner tube.

CHAPTER REVIEW

1. $L = hP$
$L = 12(8 \cdot 7)$
$L = 672 \text{ in}^2$

2. $L = hP$
$L = 11(7 + 8 + 12)$
$L = 297 \text{ cm}^2$

3.

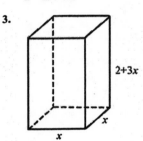

$L = hP$
$480 = (2 + 3x) \cdot 4x$
$480 = 8x + 12x^2$
$0 = 12x^2 + 8x - 480$
$0 = 3x^2 + 2x - 120$
$0 = (3x + 20)(x - 6)$
$3x + 20 = 0 \quad$ or $\quad x - 6 = 0$
$\qquad x = -\frac{20}{3} \quad$ or $\qquad x = 6$; reject $x = -\frac{20}{3}$
Dimensions are 6 in. by 6 in. by 20 in.

$V = Bh$
$V = (6 \cdot 6)(20)$
$V = 720 \text{ in}^3$

4.

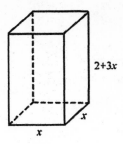

$$2+3x$$

$$x$$

$$x$$

$$L = 12(4x + 6)$$
$$360 = 48x + 72$$
$$48x = 288$$
$$x = 6$$

The dimensions of the box are $l = 9$ cm, $w = 6$ cm, $h = 12$ cm.

$$T = L + 2B$$
$$T = 360 + 2(9 \cdot 6)$$
$$T = 360 + 108$$
$$T = 468 \text{ cm}^2$$

$$V = l \cdot w \cdot h$$
$$V = 9 \cdot 6 \cdot 12$$
$$V = 648 \text{ cm}^3$$

5. The base of the prism is a right triangle since
$$15^2 = 9^2 + 12^2$$

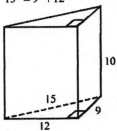

10

15

9

12

a. $L = hP$
$$L = 10(9 + 12 + 15)$$
$$L = 360 \text{ in}^2$$

b. The area of the base is
$$B = \frac{1}{2}bh$$
$$B = \frac{1}{2} \cdot 12 \cdot 9$$
$$B = 54 \text{ in}^2$$

$$T = L + 2B$$
$$T = 360 + 2 \cdot 54$$
$$T = 468 \text{ in}^2$$

c. $V = Bh$
$$V = 54 \cdot 10$$
$$V = 540 \text{ in}^3$$

6. a. $L = hP$
$$L = 13(6 \cdot 8)$$
$$L = 624 \text{ cm}^2$$

b. Using the 30°-60°-90° relationship, the apothem of the base is $4\sqrt{3}$.

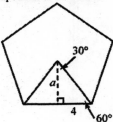

30°

a

4

60°

The area of the base is then equal to
$$B = \frac{1}{2}aP$$
$$B = \frac{1}{2} \cdot 4\sqrt{3} \cdot (6 \cdot 8)$$
$$B = 96\sqrt{3} \text{ cm}^2$$

$$T = L + 2B$$
$$T = 624 + 2 \cdot 96\sqrt{3}$$
$$T = 624 + 192\sqrt{3}$$
$$T \approx 956.55 \text{ cm}^2$$

c. $V = Bh$
$$V = \left(96\sqrt{3}\right) \cdot 13$$
$$V = 1248\sqrt{3} \approx 2161.6 \text{ cm}^3$$

7. $l^2 = a^2 + h^2$
$$l^2 = 5^2 + 8^2$$
$$l^2 = 89$$
$$l = \sqrt{89} \approx 9.43 \text{ cm}$$

8. $l^2 = a^2 + h^2$
$$12^2 = 9^2 + h^2$$
$$h^2 = 144 - 81$$
$$h^2 = 63$$
$$h = \sqrt{63} = 3\sqrt{7} \approx 7.94 \text{ in.}$$

9. $l^2 = r^2 + h^2$
$$l^2 = 5^2 + 7^2$$
$$l^2 = 74$$
$$l = \sqrt{74} \approx 8.60 \text{ in.}$$

10. $l^2 = r^2 + h^2$
$$(2r)^2 = r^2 + 6^2$$
$$4r^2 = r^2 + 36$$
$$3r^2 = 36$$
$$r^2 = 12$$
$$r = \sqrt{12} = 2\sqrt{3} \approx 3.46 \text{ cm}$$

11.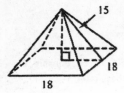

a. $L = \frac{1}{2}lP$

$L = \frac{1}{2} \cdot 15 \cdot (4 \cdot 18)$

$L = 540 \text{ in}^2$

b. $T = L + B$

$T = 540 + 18^2$

$T = 540 + 324$

$T = 864 \text{ in}^2$

c.

$l^2 = a^2 + h^2$

$15^2 = 9^2 + h^2$

$h^2 = 225 - 81$

$h^2 = 144$

$h = 12 \text{ in.}$

$V = \frac{1}{3}Bh$

$V = \frac{1}{3} \cdot 18^2 \cdot 12$

$V = 1296 \text{ in}^3$

12. a. Using the 30°-60°-90° relationship, the apothem is $a = 2\sqrt{3}$ cm.

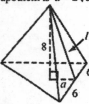

$l^2 = a^2 + h^2$

$l^2 = \left(2\sqrt{3}\right)^2 + 8^2$

$l^2 = 12 + 64$

$l^2 = 76$

$l = \sqrt{76} = 2\sqrt{19} \text{ cm}$

$L = \frac{1}{2}lP$

$L = \frac{1}{2} \cdot 2\sqrt{19} \cdot (12 \cdot 3)$

$L = 36\sqrt{19} = 156.92 \text{ cm}^2$

b. $T = L + B$

$T = 36\sqrt{19} + \frac{s^2\sqrt{3}}{4}$

$T = 36\sqrt{19} + \frac{12^2\sqrt{3}}{4}$

$T = 36\sqrt{19} + 36\sqrt{3} \approx 219.27 \text{ cm}^2$

c. $V = \frac{1}{3}Bh$

$V = \frac{1}{3}\left(36\sqrt{3}\right)(8)$

$V = 96\sqrt{3} \approx 166.28 \text{ cm}^3$

13. a. $L = hC$

$L = h \cdot 2\pi r$

$L = 10 \cdot 2\pi \cdot 6$

$L = 120\pi \text{ in}^2$

b. $T = L + 2B$

$T = 120\pi + 2\pi r^2$

$T = 120\pi + 2\pi \cdot 6^2$

$T = 120\pi + 72\pi$

$T = 192\pi \text{ in}^2$

c. $V = Bh$

$V = \pi r^2 h$

$V = \pi \cdot 6^2 \cdot 10$

$V = 360\pi \text{ in}^3$

14. a. $V = \frac{1}{2}Bh$

$V = \frac{1}{2}\pi r^2 h$

$V \approx \frac{1}{2} \cdot 3.14 \cdot 4^2 \cdot 14$

$V \approx 351.68 \text{ ft}^3$

b. The inside area and outside area represent the total area of the cylinder.

$T = L + 2B$

$T \approx 2(3.14)(4)(14) + 2(3.14) \cdot 4^2$

$T \approx 351.68 + 100.48$

$T \approx 452.16 \text{ ft}^2$

15. a.

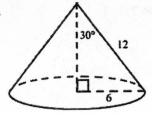

$$L = \frac{1}{2}lC$$
$$L = \frac{1}{2}l \cdot 2\pi r$$
$$L = l \cdot \pi r$$
$$L = 12 \cdot \pi \cdot 6$$
$$L = 72\pi \approx 226.19 \text{ cm}^2$$

b. $T = L + B$
$$T = 72\pi + \pi r^2$$
$$T = 72\pi + \pi \cdot 6^2$$
$$T = 72\pi + 36\pi$$
$$T = 108\pi \approx 339.29 \text{ cm}^2$$

c. $V = \frac{1}{3}Bh$
$$V = \frac{1}{3} \cdot \pi r^2 \cdot h$$
$$V = \frac{1}{3} \cdot \pi \cdot 6^2 \cdot 6\sqrt{3}$$
$$V = 72\pi\sqrt{3} \approx 391.78 \text{ cm}^3$$

16.

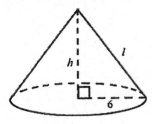

$$V = \frac{1}{3}Bh$$
$$V = \frac{1}{3}\pi r^2 \cdot h$$
$$96\pi = \frac{1}{3}\pi \cdot 6^2 \cdot h$$
$$96\pi = 12\pi h$$
$$h = \frac{96\pi}{12\pi} = 8$$

$l = 10$ using the Pythagorean Triple (6, 8, 10).

17. $S = 4\pi r^2$
$$S \approx 4 \cdot \frac{22}{7} \cdot 7^2$$
$$S \approx 616 \text{ in}^2$$

18. $V = \frac{4}{3}\pi r^3$
$$V \approx \frac{4}{3}(3.14) \cdot 6^3$$
$$V \approx 904.32 \text{ cm}^3$$

19.

$$V = V_{\text{HEMISPHERE}} + V_{\text{CYLINDER}} + V_{\text{CONE}}$$
$$V = \frac{1}{2}\left(\frac{4}{3}\pi r^3\right) + Bh + \frac{1}{3}Bh$$
$$V = \frac{1}{2}\left(\frac{4}{3}\pi r^3\right) + \pi r^2 h + \frac{1}{3}\pi r^2 h$$
$$V = \frac{1}{2}\left(\frac{4}{3}\pi \cdot 3^3\right) + \pi \cdot 3^2 \cdot 10 + \frac{1}{3}\pi \cdot 3^2 \cdot 4$$
$$V = 18\pi + 90\pi + 12\pi$$
$$V = 120\pi \text{ units}^3$$

20. Let r equal the radius of the smaller sphere while $3r$ is the radius of the larger sphere.

$$\frac{\text{Surface Area of Smaller}}{\text{Surface Area of Larger}} = \frac{4\pi r^2}{4\pi (3r)^2} = \frac{r^2}{9r^2} = \frac{1}{2}$$

$$\frac{\text{Volume of Smaller}}{\text{Volume of Larger}} = \frac{\frac{4}{3}\pi r^3}{\frac{4}{3}\pi (3r)^3} = \frac{r^3}{27r^3} = \frac{1}{27}$$

21. The solid formed is a cone.
$$V = \frac{1}{3}Bh$$
$$V = \frac{1}{3} \cdot \pi r^2 \cdot h$$
$$V \approx \frac{1}{3} \cdot \frac{22}{7} \cdot 5^2 \cdot 7$$
$$V \approx 183\frac{1}{3} \text{ in}^3$$

22. The solid formed is a cylinder.
$$V = Bh$$
$$V = \pi r^2 h$$
$$V = \pi \cdot 6^2 \cdot 8$$
$$V = 288\pi \text{ cm}^3$$

23. The solid formed is a sphere.
$$V = \frac{4}{3}\pi r^3$$
$$V = \frac{4}{3}\pi \cdot 2^3$$
$$V = \frac{32\pi}{3} \text{ in}^3$$

24.

$$V = V_{\text{LARGER CYLINDER}} - V_{\text{SMALLER CYLINDER}}$$
$$V = \pi R^2 h - \pi r^2 h$$
$$V \approx (3.14)(5^2)(36) - (3.14)(4^2)(36)$$
$$V \approx 2826 - 1808.64$$
$$V \approx 1017.36 \text{ in}^3$$

25. $V = V_{CUBE} - V_{SPHERE}$

$V = e^3 - \frac{4}{3}\pi r^3$

$V = 14^3 - \frac{4}{3}\pi \cdot 7^3$

$V = 2744 - \frac{1372\pi}{3}$ in^3

26. a. An octahedron has eight faces which are equilateral triangles.

b. A tetrahedron has four faces which are equilateral triangles.

c. A dodecahedron has twelve faces which are regular pentagons.

27. $V = V_{CYLINDER} + V_{2\ HEMISPHERES} - V_{SPHERE}$

Because the volume of the 2 hemispheres equals the volume of a sphere, we have

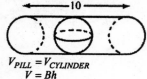

$V_{PILL} = V_{CYLINDER}$

$V = Bh$

$V = \pi r^2 h$

$V = \pi \cdot 2^2 \cdot 10$

$V = 40\pi$ mm^3

28. a. $V = 16$, $E = 24$, $F = 10$

$V + F = E + 2$

$16 + 10 = 24 + 2$

b. $V = 4$, $E = 6$, $F = 4$

$V + F = E + 2$

$4 + 4 = 6 + 2$

c. $V = 6$, $E = 12$, $F = 8$

$V + F = E + 2$

$6 + 8 = 12 + 2$

29. $V = 10 \cdot 3 \cdot 4 - 2 \cdot (1 \cdot 1 \cdot 3)$

$V = 120 - 6$

$V = 114$ in^3

30. a. $\frac{4}{8} = \frac{1}{2}$

b. $\frac{5}{8}$

31. a. $A = 6.5(12) = 78$ in^2

b. $A = 4\left(\frac{s^2\sqrt{3}}{4}\right) = 4\left(\frac{4^2\sqrt{3}}{4}\right) = 16\sqrt{3}$ cm^2

CHAPTER TEST

1. a. 15

b. 7

2. a. $P = 5 \cdot 3.2 = 16$ cm

$A = \frac{1}{2}aP$

$= \frac{1}{2}(2 \cdot 16)$

$= 16$ cm^2

b. $L = hP = 5 \cdot 16 = 80$

$T = L + 2B$

$T = 80 + 2\left(\frac{1}{2}aP\right)$

$T = 80 + 2(16)$

$T = 112$ cm^2

c. $V = Bh$

$V = \left(\frac{1}{2}aP\right)(5)$

$V = 16 \cdot 5$

$V = 80$ cm^3

3. a. 5

b. 4

4.

$6^2 = 2^2 + l^2$

$l^2 = 32$

$l = \sqrt{32} = 4\sqrt{2}$ ft

a. $L = \frac{1}{2}lP$

$L = \frac{1}{2}(4\sqrt{2})(16)$

$L = 32\sqrt{2}$ ft^2

b. $T = B + L$

$T = 16 + 32\sqrt{2} \approx 61.25$ ft^2

5.

$17^2 = 8^2 + l^2$

$l^2 = 225$

$l = 15$ ft

6.

$5^2 = 4^2 + l^2$
$l^2 = 9$
$l = 3$ in.

7. $V = \frac{1}{3}Bh$

$V = \frac{1}{3}(5 \cdot 5)(6)$

$V = 50$ ft^3

8. **a.** False

b. True

9. **a.** True

b. True

10. $V = 6$, $F = 8$
$V + F = E + 2$
$6 + 8 = E + 2$
$E = 12$

11.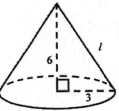

$l^2 = 3^2 + 6^2$
$l^2 = 45$
$l = \sqrt{45} = 3\sqrt{5}$ cm

12. **a.** $L = 2\pi rh$
$L = 2\pi \cdot 4 \cdot 6$
$L = 48\pi$ cm^2

b. $V = \pi r^2 h$
$V = \pi \cdot 4^2 \cdot 6$
$V = 96\pi$ cm^3

13.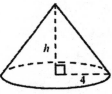

$V = \frac{1}{3}\pi r^2 h$

$32\pi = \frac{1}{3}\pi \cdot 4^2 \cdot h$

$h = 6$ in.

14. **a.** $\frac{4}{8} = \frac{1}{2}$

b. $\frac{3}{8}$

15. **a.** $S = 4\pi r^2$
$S = 4\pi \cdot 10^2$
$S = 400\pi \approx 1256.6$ ft^2

b. $V = \frac{4}{3}\pi r^3$
$V = \frac{4}{3}\pi \cdot 10^3$
$V = \frac{4000}{3}\pi \approx 4188.8$ ft^3

Chapter 9: Analytic Geometry

SECTION 9.1: The Rectangular Coordinate System

1.

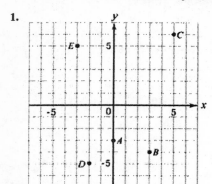

5. The segment is horizontal so $d = 7 - b$ if $7 > b$
$$\therefore\ 7 - b = 3.5$$
$$-b = -3.5$$
$$b = 3.5$$
If $b > 7$, then $d = b - 7$.
$$\therefore\ b - 7 = 3.5$$
$$b = 10.5$$

9. a. $M = \left(\dfrac{x_1 + x_2}{2}, \dfrac{y_1 + y_2}{2}\right)$
$$M = \left(\dfrac{0 + 4}{2}, \dfrac{(-3) + 0}{2}\right)$$
$$M = \left(\dfrac{4}{2}, -\dfrac{3}{2}\right) = \left(2, -\dfrac{3}{2}\right)$$

b. $M = \left(\dfrac{(-2) + 4}{2}, \dfrac{5 + (-3)}{2}\right)$
$$M = \left(\dfrac{2}{2}, \dfrac{3}{2}\right)$$
$$M = (1, 1)$$

c. $M = \left(\dfrac{3 + 5}{2}, \dfrac{2 + (-2)}{2}\right)$
$$M = \left(\dfrac{8}{2}, \dfrac{0}{2}\right)$$
$$M = (4, 0)$$

d. $M = \left(\dfrac{a + 0}{2}, \dfrac{0 + b}{2}\right)$
$$M = \left(\dfrac{a}{2}, \dfrac{b}{2}\right)$$

13. a. $\left(4, -\dfrac{5}{2}\right)$

b. $(0, 4)$

c. $\left(\dfrac{7}{2}, -1\right)$

d. (a, b)

17. a. $(-3, -4)$

b. $(-2, 0)$

c. $(-a, 0)$

d. $(-b, c)$

21. $M = \left(\dfrac{x_1 + x_2}{2}, \dfrac{y_1 + y_2}{2}\right)$
$$(2.1, -5.7) = \left(\dfrac{x + 1.7}{2}, \dfrac{y + 2.3}{2}\right)$$
$$\dfrac{x + 1.7}{2} = 2.1 \quad \text{and} \quad \dfrac{y + 2.3}{2} = -5.7$$
$$x + 1.7 = 4.2 \quad \text{and} \quad y + 2.3 = -11.4$$
$$x = 2.5 \quad \text{and} \qquad y = -13.7$$
$$B = (2.5, -13.7)$$

25. a. $AB = 4 - 0 = 4$
$$BC = \sqrt{(2 - 4)^2 + (5 - 0)^2}$$
$$= \sqrt{(-2)^2 + 5^2}$$
$$= \sqrt{4 + 25} = \sqrt{29}$$
$$AC = \sqrt{(2 - 0)^2 + (5 - 0)^2}$$
$$= \sqrt{2^2 + 5^2}$$
$$= \sqrt{4 + 25} = \sqrt{29}$$
Because $BC = AC$, $\triangle ABC$ is isosceles.

b. $DE = 4 - 0 = 4$
$$DF = \sqrt{(2 - 0)^2 + \left(2\sqrt{3} - 0\right)^2}$$
$$= \sqrt{2^2 + \left(2\sqrt{3}\right)^2}$$
$$= \sqrt{4 + 12} = \sqrt{16} = 4$$
$$EF = \sqrt{(2 - 4)^2 + \left(2\sqrt{3} - 0\right)^2}$$
$$= \sqrt{(-2)^2 + \left(2\sqrt{3}\right)^2}$$
$$= \sqrt{4 + 12} = \sqrt{16} = 4$$
Because $DE = DF = EF$, $\triangle DEF$ is equilateral.

192

c. $GH = \sqrt{(-2-[-5])^2 + (6-2)^2}$
$= \sqrt{3^2 + 4^2}$
$= \sqrt{9+16} = \sqrt{25} = 5$

$GK = \sqrt{(2-[-5])^2 + (3-2)^2}$
$= \sqrt{7^2 + 1^2}$
$= \sqrt{49+1} = \sqrt{50} = 5\sqrt{2}$

$HK = \sqrt{(2-[-2])^2 + (3-6)^2}$
$= \sqrt{4^2 + (-3)^2}$
$= \sqrt{16+9} = \sqrt{25} = 5$

Because $GH = HK$ and
$(GH)^2 + (HK)^2 = (GK)^2$, $\triangle GHK$ is an
isosceles right triangle.

29. Call the third vertex (a, b). Because the three
sides must be of the same length and $AB = 2a$,
we have
$$\sqrt{(a-0)^2 + (b-0)^2} = 2a$$
$$\sqrt{a^2 + b^2} = 2a$$
Squaring, $a^2 + b^2 = 4a^2$
$$b^2 = 3a^2$$
$$b = \pm\sqrt{3a^2}$$
$$b = \pm a\sqrt{3}$$

The third vertex is $(a, a\sqrt{3})$ or $(a, -a\sqrt{3})$.

33. $A_{MNQ} = A_{\text{RECT}} - (A_{\triangle 1} + A_{\triangle 2} + A_{\triangle 3})$
$$A = (7)(5) - \left[\frac{1}{2}(1)(5) + \frac{1}{2}(6)(4) + \frac{1}{2}(7)(1)\right]$$
$$A = 35 - \left[\frac{5}{2} + 12 + \frac{7}{2}\right]$$
$$A = 35 - 18$$
$$A = 17$$

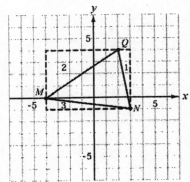

37. a. A cone is formed; $r = 9, h = 5$
$$V = \frac{1}{3}\pi r^2 h$$
$$V = \frac{1}{3}\pi \cdot 9^2 \cdot 5$$
$$V = 135\pi \text{ units}^2$$

b. A cone is formed; $r = 5, h = 9$
$$V = \frac{1}{3}\pi r^2 h$$
$$V = \frac{1}{3}\pi \cdot 5^2 \cdot 9$$
$$V = 75\pi \text{ units}^2$$

41. a. $L = 2\pi rh$
$L = 2\pi \cdot 5 \cdot 9$
$L = 90\pi \text{ units}^2$

b. $L = 2\pi rh$
$L = 2\pi \cdot 9 \cdot 5$
$L = 90\pi \text{ units}^2$

45. a. $(5, 4)$

b. $(1, 8)$

c. $(3, 2)$

SECTION 9.2: Graphs of Linear Equations and Slope

1. $3x + 4y = 12$ has intercepts (4, 0) and (0, 3).

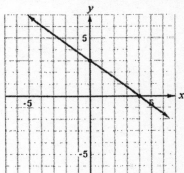

5. $2x + 6 = 0$ is equivalent to $x = -3$. It is a vertical
line with x-intercept $(-3, 0)$.

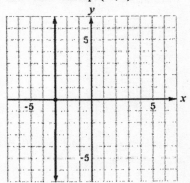

9. a. $m = \dfrac{y_2 - y_1}{x_2 - x_1}$

$m = \dfrac{5 - (-3)}{4 - 2}$

$m = \dfrac{8}{2}$

$m = 4$

b. $m = \dfrac{7 - (-2)}{3 - 3} = \dfrac{9}{0}$

m is undefined.

c. $m = \dfrac{-2 - (-1)}{2 - 1}$

$m = \dfrac{-1}{1}$

$m = -1$

d. $m = \dfrac{5 - 5}{(-1.3) - (-2.7)}$

$m = \dfrac{0}{1.4}$

$m = 0$

e. $m = \dfrac{d - b}{c - a}$

f. $m = \dfrac{b - 0}{0 - a}$

$m = -\dfrac{b}{a}$

13. a. $m_{\overline{AB}} = \dfrac{2 - 5}{0 - (-2)}$

$m_{\overline{AB}} = -\dfrac{3}{2}$

$m_{\overline{BC}} = \dfrac{-4 - 2}{4 - 0}$

$m_{\overline{BC}} = \dfrac{-6}{4}$

$m_{\overline{BC}} = -\dfrac{3}{2}$

Because $m_{\overline{AB}} = m_{\overline{BC}}$, the points A, B and C are collinear.

b. $m_{\overline{DE}} = \dfrac{-2 - (-1)}{2 - (-1)}$

$m_{\overline{DE}} = -\dfrac{1}{3}$

$m_{\overline{EF}} = \dfrac{-5 - (-2)}{5 - 2}$

$m_{\overline{EF}} = \dfrac{-3}{3}$

$m_{\overline{EF}} = -1$

Because $m_{\overline{DE}} \neq m_{\overline{EF}}$, the points D, E and F are noncollinear.

17. a. 2

b. $-\dfrac{4}{3}$

c. $-\dfrac{1}{3}$

d. $-\dfrac{h + j}{f + g}$

21. $2x + 3y = 6$ contains $(3, 0)$ and $(0, 2)$ so that its slope is $m_1 = \dfrac{2 - 0}{0 - 3} = -\dfrac{2}{3}$.

$3x - 2y = 12$ contains $(4, 0)$ and $(0, -6)$ so that its slope is $m_2 = \dfrac{-6 - 0}{0 - 4} = \dfrac{-6}{-4} = \dfrac{3}{2}$.

Because $m_1 \cdot m_2 = -1$, these lines are perpendicular.

25. The 2 lines are perpendicular if $m_1 \cdot m_2 = -1$.

$m_1 = \dfrac{2 - (-3)}{3 - 2} = \dfrac{5}{1} = 5$

$m_2 = \dfrac{-1 - 4}{x - (-2)} = \dfrac{-5}{x + 2}$

$\dfrac{5}{1} \cdot \dfrac{-5}{x + 2} = -1$

$\dfrac{-25}{x + 2} = -1$

$-25 = -1(x + 2)$

$-25 = -1x - 2$

$-23 = -1x$

$x = 23$

29. First plot $(0, 5)$. If $m = -\dfrac{3}{4}$, then $m = \dfrac{-3}{4}$; a change in y of -3 corresponds to a change in x of 4.

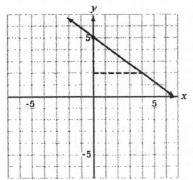

33. Let $A = (6, 5)$, $B = (-3, 0)$, and $C = (4, -2)$.

$m_{\overline{AB}} = \dfrac{0 - 5}{-3 - 6} = \dfrac{-5}{-9} = \dfrac{5}{9}$

$m_{\overline{BC}} = \dfrac{-2 - 0}{4 - (-3)} = \dfrac{-2}{7}$

$m_{\overline{AC}} = \dfrac{-2 - 5}{4 - 6} = \dfrac{-7}{-2} = \dfrac{7}{2}$

Because $m_{\overline{BC}} \cdot m_{\overline{AC}} = -1$, $\overline{BC} \perp \overline{AC}$ and $\triangle ABC$ is a right triangle.

37. $m_{\overline{VT}} = \dfrac{e-e}{(c-d)-(a+d)} = \dfrac{0}{c-a-2d} = 0$

$m_{\overline{RS}} = \dfrac{b-b}{c-a} = \dfrac{0}{c-a} = 0$

$\therefore \overline{VT} \parallel \overline{RS}$

$RV = \sqrt{[(a+d)-a]^2 + (e-b)^2}$
$\quad = \sqrt{d^2 + (e-b)^2}$
$\quad = \sqrt{d^2 + e^2 - 2be + b^2}$

$ST = \sqrt{[c-(c-d)]^2 + (b-e)^2}$
$\quad = \sqrt{d^2 + (b-e)^2}$
$\quad = \sqrt{d^2 + b^2 - 2be + e^2}$

$\therefore \overline{RV} \parallel \overline{ST}$

RSTV is an isosceles trapezoid.

41. If $l_1 \parallel l_2$, then $\angle A \cong \angle D$.

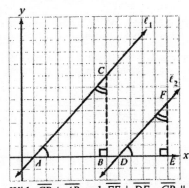

With $\overline{CB} \perp \overline{AB}$ and $\overline{FE} \perp \overline{DE}$, $\overline{CB} \parallel \overline{FE}$ (both $\perp$ to the *x*-axis). Then $\angle B \cong \angle E$ (all rt. $\angle$s are $\cong$). By AA, $\triangle ABC \sim \triangle DEF$.

Then $\dfrac{CB}{FE} = \dfrac{AB}{DE}$ since corr. sides of $\sim \triangle$s are proportional. By a property of proportions, $\dfrac{CB}{AB} = \dfrac{FE}{DE}$ (means were interchanged).

But $m_1 = \dfrac{CB}{AB}$ and $m_2 = \dfrac{FE}{DE}$.

Then $m_1 = m_2$ and the slopes are equal.

SECTION 9.3: Preparing to Do Analytic Proofs

1. a. $d = \sqrt{(x_2 - x_1)^2 + (y_2 - y_1)^2}$
$d = \sqrt{(0-a)^2 + (a-0)^2}$
$d = \sqrt{(-a)^2 + a^2}$
$d = \sqrt{a^2 + a^2}$
$d = \sqrt{2a^2} = a\sqrt{2}$ if $a > 0$

b. $m = \dfrac{y_2 - y_1}{x_2 - x_1}$

$m = \dfrac{d-b}{c-a}$

5. $\overline{AB}$ is horizontal and $\overline{BC}$ is vertical.
$\therefore \overline{AB} \perp \overline{BC}$. Hence $\angle B$ is a right $\angle$ and $\triangle ABC$ is a right triangle.

9. $m_{\overline{MN}} = 0$ and $m_{\overline{QP}} = 0$. $\therefore \overline{MN} \parallel \overline{QP}$

$\overline{QM}$ and $\overline{PN}$ are both vertical; $\therefore \overline{QM} \parallel \overline{PN}$. Hence, *MQPN* is a parallelogram.

Since $\overline{QM}$ is vertical and $\overline{MN}$ is horizontal, $\angle QMN$ is a right angle. Because parallelogram *MQPN* has a right $\angle$, it is also a rectangle.

13. $M = (0, 0)$; $N = (r, 0)$; $P = (r + s, t)$

17. a. Square

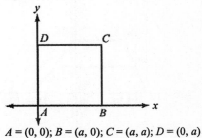

$A = (0, 0)$; $B = (a, 0)$; $C = (a, a)$; $D = (0, a)$

b. Square (with midpoints of sides)
$A = (0, 0)$; $B = (2a, 0)$
$C = (2a, 2a)$; $D = (0, 2a)$

21. a. Isosceles triangle

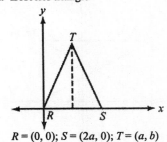

$R = (0, 0)$; $S = (2a, 0)$; $T = (a, b)$

b. Isosceles triangle (with midpoints)
$R = (0, 0)$; $S = (4a, 0)$; $T = (2a, 2b)$

25. Parallelogram $ABCD$ with $\overline{AC} \perp \overline{DB}$

$$m_{\overline{AC}} = \frac{c-0}{a+b-0} = \frac{c}{a+b}$$

$$m_{\overline{DB}} = \frac{0-c}{a-b} = \frac{-c}{a-b}$$

$$\therefore \frac{c}{a+b} \cdot \frac{-c}{a-b} = -1$$

$$\frac{-c^2}{a^2-b^2} = -1$$

$$-c^2 = -(a^2-b^2)$$

$$c^2 = a^2 - b^2$$

29. **a.** a is positive

   b. $-a$ is negative

   c. $AB = a - (-a) = 2a$

33. The line segment joining the midpoints of the two nonparallel sides of a trapezoid is parallel to the bases of the trapezoid.

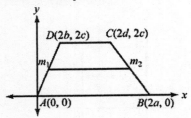

SECTION 9.4: Analytic Proofs

1. The diagonals of a rectangle are equal in length.
   Proof: Let rectangle $ABCD$ have vertices as shown.

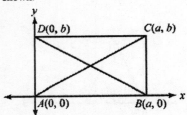

Then $AC = \sqrt{(a-0)^2 + (b-0)^2}$
$$= \sqrt{a^2 + b^2}$$

Also $DB = \sqrt{(a-0)^2 + (0-b)^2}$
$$= \sqrt{a^2 + (-b)^2} = \sqrt{a^2 + b^2}$$

Then $AC = DB$, and the diagonals of the rectangle are equal in length.

5. The median from the vertex of an isosceles triangle to the base is perpendicular to the base.

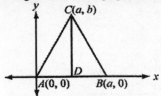

Proof: The triangle ABC with vertices as shown is isosceles because $AC = BC$.

Let D be the midpoint of $\overline{AB}$.

$$D = \left(\frac{0+2a}{2}, \frac{0+0}{2}\right) = (a, 0).$$

$m_{\overline{AB}} = \frac{0-0}{2a-0} = 0$; that is $\overline{AB}$ is horizontal.

$m_{\overline{CD}} = \frac{b-0}{a-a} = \frac{b}{0}$ which is undefined; that is $\overline{CD}$ is vertical.

Then $\overline{CD} \perp \overline{AB}$ and the median to the base of the isosceles triangle is perpendicular to the base.

9. The segments which join the midpoints of the consecutive sides of a rectangle form a rhombus.

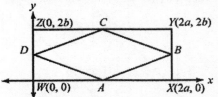

Proof: With vertices as shown, $WXYZ$ is a rectangle.
Midpoints of the sides of the rect. Are

$$A = \left(\frac{0+2a}{2}, \frac{0+0}{2}\right) = (a, 0)$$

$$B = \left(\frac{2a+2a}{2}, \frac{0+2b}{2}\right) = (2a, b)$$

$$C = \left(\frac{0+2a}{2}, \frac{2b+2b}{2}\right) = (a, 2b)$$

$$D = \left(\frac{0+0}{2}, \frac{0+2b}{2}\right) = (0, b)$$

$m_{\overline{AB}} = \frac{b-0}{2a-a} = \frac{b}{a}$ and

$m_{\overline{DC}} = \frac{b-a}{0-a} = \frac{-b}{-a} = \frac{b}{a}$, so $\overline{AB} \parallel \overline{DC}$.

$m_{\overline{DA}} = \frac{b-0}{0-a} = \frac{b}{-a}$ and

$m_{\overline{CB}} = \frac{2b-b}{a-2a} = \frac{-b}{-a} = \frac{b}{a}$, so $\overline{DA} \parallel \overline{CB}$.

Then $ABCD$ is a parallelogram.
We need to show that 2 adjacent sides are congruent (equal in length).

$$AB = \sqrt{(2a-a)^2 + (b-0)^2} = \sqrt{a^2 + b^2}$$

$$BC = \sqrt{(a-2a)^2 + (2b-b)^2}$$
$$= \sqrt{(-a)^2 + b^2} = \sqrt{a^2 + b^2}$$

13. The segment which joins the midpoints of 2 sides of a triangle is parallel to the third side and has a length equal to one-half the length of the third side.

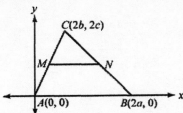

Proof: Let $\triangle ABC$ have vertices as shown. With M and N the midpoints of $\overline{AC}$ and $\overline{BC}$ respectively,

$$M = \left(\frac{0+2b}{2}, \frac{0+2c}{2}\right) = (b, c) \text{ and}$$

$$N = \left(\frac{2a+2b}{2}, \frac{0+2c}{2}\right) = (a+b, c).$$

Now $m_{\overline{MN}} = \dfrac{c-c}{(a+b)-b} = \dfrac{0}{a} = 0$ and

$$m_{\overline{AB}} = \frac{0-0}{2a-0} = \frac{0}{2a} = 0$$

Then $\overline{MN} \parallel \overline{AB}$.

Also $MN = (a+b) - b = a$
$\qquad AB = 2a - 0 = 2a$

Then $MN = \dfrac{1}{2}(AB)$

That is, the segment $\left(\overline{MN}\right)$ which joins the midpoints of 2 sides of the triangle is parallel to the third side and equals one-half its length.

17. If the diagonals of a parallelogram are perpendicular, then the parallelogram is a rhombus.

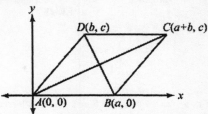

Proof: Let parallelogram $ABCD$ have vertices as shown. If $\overline{AC} \perp \overline{DB}$, then $m_{\overline{AC}} \cdot m_{\overline{DB}} = -1$.

Because $m_{\overline{AC}} = \dfrac{c-0}{(a+b)-0} = \dfrac{c}{a+b}$ and

$m_{\overline{DB}} = \dfrac{0-c}{a-b} = \dfrac{-c}{a-b}$, it follows that

$$\frac{c}{a+b} \cdot \frac{-c}{a-b} = -1$$

$$\frac{-c^2}{a^2 - b^2} = -1$$

$$-c^2 = -1(a^2 - b^2)$$

$$-c^2 = -a^2 + b^2$$

$$a^2 = b^2 + c^2 (*)$$

For $ABCD$ to be a rhombus, we must show that two adjacent sides are congruent. $AB = a - 0 = a$ and $AD = \sqrt{(b-0)^2 + (c-0)^2} = \sqrt{b^2 + c^2}$.

Because $a^2 = b^2 + c^2 (*)$, it follows that

$a = \sqrt{b^2 + c^2}$ and $AB = AD$. Then parallelogram $ABCD$ is a rhombus.

21. $Ax + By = C$
$\qquad By = -Ax + C$
$$\qquad y = -\frac{A}{B}x + \frac{C}{B}$$

So $m_1 = -\dfrac{A}{B}$

$Bx - Ay = D$
$\qquad -Ay = -Bx + D$
$$\qquad y = \frac{B}{A}x - \frac{D}{A}$$

So $m_2 = \dfrac{B}{A}$

Since $m_1 \cdot m_2 = -\dfrac{A}{B} \cdot \dfrac{B}{A} = -1$, then the lines are perpendicular.

25. $x^2 + y^2 = r^2$
$\qquad x^2 + y^2 = 3^2$
$\qquad x^2 + y^2 = 9$

SECTION 9.5: Equations of Lines

1. Dividing by 8, $8x + 16y = 48$ becomes
   $x + 2y = 6$.
   Then $2y = -1x + 6$
   $$\frac{1}{2}(2y) = \frac{1}{2}(-1x + 6)$$
   $$y = -\frac{1}{2}x + 3$$

5. $y = 2x - 3$ has $m = 2$ and $b = -3$. Plot the point
   $(0, -3)$. Then draw the line for which an increase
   of 2 in y corresponds to an increase of 1 in x.

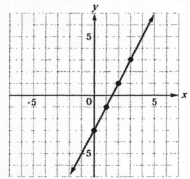

9. Using $y = mx + b$,
   $$y = -\frac{2}{3}x + 5$$
   $$3y = -2x + 15$$
   $$2x + 3y = 15$$

13. $m = \dfrac{1 - (-1)}{3 - 0} = \dfrac{2}{3}$
    Using $y = mx + b$, $y = \dfrac{2}{3}x - 1$.
    Mult. by 3,
    $$3y = 2x - 3$$
    $$-2x + 3y = -3$$

17. The graph contains $(2, 0)$ and $(0, -2)$.
    $m = \dfrac{-2 - 0}{0 - 2} = \dfrac{-2}{-2} = 1$.
    Using $y = mx + b$,
    $$y = 1x + 2$$
    $$-x + y = -2$$

21. The line $y = \dfrac{3}{4}x - 5$ has slope $m_1 = \dfrac{3}{4}$. The
    desired line has $m_2 = -\dfrac{4}{3}$. Using $y = mx + b$,
    $y = -\dfrac{4}{3}x - 4$. Mult. by 3,
    $$3y = -4x - 12$$
    $$4x + 3y = -12$$

25. Perpendicular slope is $-\dfrac{b}{a}$.
    $$y - h = -\frac{b}{a}(x - g)$$
    $$y = h = -\frac{b}{a}x + \frac{bg}{a}$$
    $$y = -\frac{b}{a}x + \frac{bg + ha}{a}$$

29. $2x + y = 6$
    $y = -2x + 6$
    $3x - y = 19$
    $y = 3x - 19$

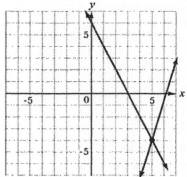

The lines intersect at (5, –4).

33. $2x + y = 8$
    $\underline{3x - y = 7}$
    $5x = 15$
    $x = 3$
    $2(3) + y = 8$
    $6 + y = 8$
    $y = 2$

37. $2x + 3y = 4$ Multiply by -3
      $3x - 4y = 23$ Multiply by 2

      $-6x - 9y = -12$
      $\underline{6x - 8y = 46}$
      $-17y = 34$
      $y = -2$
      $2x + 3(-2) = 4$
      $2x - 6 = 4$
      $2x = 10$
      $x = 5$
      The point of intersection is (5, -2).

41. $bx + c = a$
    $bx = a - c$
    $x = \dfrac{a - c}{b}$
    and $y = a$.
    $\left(\dfrac{a - c}{b}, a\right)$

45. The altitudes of a triangle are concurrent.

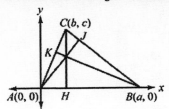

Proof: For $\triangle ABC$, let $\overline{CH}$, $\overline{AJ}$, and $\overline{BK}$ name the altitudes. Because $\overline{AB}$ is horizontal $\left(m_{\overline{AB}} = 0\right)$, $\overline{CH}$ is vertical and has the equation $x = b$. Because $m_{\overline{BC}} = \dfrac{c-0}{b-a} = \dfrac{c}{b-a}$, the slope of altitude $\overline{AJ}$ is $m_{\overline{AJ}} = -\dfrac{b-a}{c} = \dfrac{a-b}{c}$. Since $\overline{AJ}$ contains $(0, 0)$, its equation is $y = \dfrac{a-b}{c}x$.

The intersection of altitudes $\overline{CH}$ ($x = b$) and $\overline{AJ}$ $\left(y = \dfrac{a-b}{c}x\right)$ is at $x = b$ so that

$y = \dfrac{a-b}{c} \cdot b = \dfrac{b(a-b)}{c} = \dfrac{ab-b^2}{c}$. That is, $\overline{CH}$ and $\overline{AJ}$ intersect at $\left(b, \dfrac{ab-b^2}{c}\right)$. The remaining altitude is $\overline{BK}$. Since $m_{\overline{AC}} = \dfrac{c-0}{b-0} = \dfrac{c}{b}$,

$m_{\overline{BK}} = -\dfrac{b}{c}$. Because $\overline{BK}$ contains $(a, 0)$, its equation is $y - 0 = -\dfrac{b}{c}(x-a)$ or $y = \dfrac{-b}{c}(x-a)$.

For the three altitudes to be concurrent, $\left(b, \dfrac{ab-b^2}{c}\right)$ must lie on the line $y = \dfrac{-b}{c}(x-a)$.
Substitution leads to

$\dfrac{ab-b^2}{c} = \dfrac{-b}{c}(b-a)$

$\dfrac{ab-b^2}{c} = \dfrac{-b(b-a)}{c}$

$\dfrac{ab-b^2}{c} = \dfrac{-b^2+ab}{c}$, which is true.

Therefore, the three altitudes are concurrent.

CHAPTER REVIEW

1. a. $d = y_2 - y_1$ (if $y_2 > y_1$) $= 4 - (-3) = 7$

   b. $d = x_2 - x_1$ (if $x_2 > x_1$) $= 1 - (-5) = 6$

   c. $d = \sqrt{(x_2 - x_1)^2 + (y_2 - y_1)^2}$

       $d = \sqrt{(7-(-5))^2 + (-3-2)^2}$

       $d = \sqrt{12^2 + (-5)^2} = \sqrt{144+25} = \sqrt{169} = 13$

   d. $d = \sqrt{(x-(x-3))^2 + ((y-2)-(y+2))^2}$

       $d = \sqrt{3^2 + (-4)^2} = \sqrt{9+16} = \sqrt{25} = 5$

2. a. $d = y_2 - y_1$ (if $y_2 > y_1$) $= 5 - (-3) = 8$

   b. $d = x_2 - x_1$ (if $x_2 > x_1$) $= 3 - (-7) = 10$

   c. $d = \sqrt{(4-(-4))^2 + (5-1)^2}$

       $d = \sqrt{8^2 + 4^2} = \sqrt{64+16} = \sqrt{80} = 4\sqrt{5}$

   d. $d = \sqrt{((x+4)-(x-2))^2 + ((y+5)-(y-3))^2}$

       $d = \sqrt{6^2 + 8^2} = \sqrt{36+64} = \sqrt{100} = 10$

3. a. $M = \left(\dfrac{x_1 + x_2}{2}, \dfrac{y_1 + y_2}{2}\right)$

       $M = \left(\dfrac{6+6}{2}, \dfrac{4+(-3)}{2}\right) = \left(6, \dfrac{1}{2}\right)$

   b. $M = \left(\dfrac{1+(-5)}{2}, \dfrac{4+4}{2}\right) = (-2, 4)$

   c. $M = \left(\dfrac{(-5)+7}{2}, \dfrac{2+(-3)}{2}\right) = \left(1, -\dfrac{1}{2}\right)$

   d. $M = \left(\dfrac{(x-3)+x}{2}, \dfrac{(y+2)+(y-2)}{2}\right)$

       $M = \left(\dfrac{2x-3}{2}, y\right)$

4. a. $M = \left(\dfrac{2+2}{2}, \dfrac{(-3)+5}{2}\right) = (2, 1)$

   b. $M = \left(\dfrac{3+(-7)}{2}, \dfrac{(-2)+(-2)}{2}\right) = (-2, -2)$

   c. $M = \left(\dfrac{(-4)+4}{2}, \dfrac{1+5}{2}\right) = (0, 3)$

   d. $M = \left(\dfrac{(x-2)+(x+4)}{2}, \dfrac{(y-3)+(y+5)}{2}\right)$

       $M = (x+1, y+1)$

5. **a.** $m = \dfrac{y_2 - y_1}{x_2 - x_1}$

        $m = \dfrac{(-3)-4}{6-6} = \dfrac{-7}{0}$ m is undefined.

    b. $m = \dfrac{4-4}{-5-1} = \dfrac{0}{-6} = 0$

    c. $m = \dfrac{-3-2}{7-(-5)} = \dfrac{-5}{12}$

    d. $m = \dfrac{(y-2)-(y+2)}{x-(x-3)} = \dfrac{-4}{3}$

6. **a.** $m = \dfrac{5-(-3)}{2-2} = \dfrac{8}{0}$ m is undefined.

    b. $m = \dfrac{-2-(-2)}{-7-3} = \dfrac{0}{-10} = 0$

    c. $m = \dfrac{5-1}{4-(-4)} = \dfrac{4}{8} = \dfrac{1}{2}$

    d. $m = \dfrac{(y+5)-(y-3)}{(x+4)-(x-2)} = \dfrac{8}{6} = \dfrac{4}{3}$

7. $M = \left(\dfrac{x_1 + x_2}{2}, \dfrac{y_1 + y_2}{2} \right)$

     $(2,1) = \left(\dfrac{8+x}{2}, \dfrac{10+y}{2} \right)$

     $\dfrac{8+x}{2} = 2$ and $\dfrac{10+y}{2} = 1$

     $8 + x = 4$ and $10 + y = 2$

      $x = -4$ and $y = -8$

     $B = (-4, \ -8)$

8. Due to symmetry, $R = (3, 7)$.

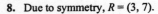
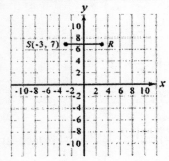

9. $m = \dfrac{y_2 - y_1}{x_2 - x_1}$

     $-3 = \dfrac{3-1}{x-2}$

     $-3 = \dfrac{2}{x-2}$

   $-3(x-2) = 2$

    $-3x + 6 = 2$

      $-3x = -4$

       $x = \dfrac{4}{3}$

10. $m = \dfrac{y_2 - y_1}{x_2 - x_1}$

     $\dfrac{-6}{7} = \dfrac{y-2}{2-(-5)}$

     $\dfrac{-6}{7} = \dfrac{y-2}{7}$

     $-6 = y - 2$

      $y = -4$

11. **a.** $x + 3y = 6$

       $y = -\dfrac{1}{3}x + 2$

       $m = -\dfrac{1}{3}$

     $3x - y = -7$

       $y = 3x + 7$

       $m = 3$

    Since $m_1 \cdot m_2 = -1$, the lines are perpendicular.

    b. $2x - y = -3$

       $y = 2x + 3$

       $m = 2$ and $b = 3$

      $y = 2x - 14$

      $m = 2$ and $b = -14$

    Since $m_1 = m_2$, the lines are parallel.

    c. $y + 2 = -3(x - 5)$

       $y = -3x + 13$

       $m = -3$

     $2y = 6x + 11$

       $y = 3x + \dfrac{11}{2}$

       $m = 3$

    The lines are neither parallel nor perpendicular.

    d. $0.5x + y = 0$

       $y = -\dfrac{1}{2}x$

       $m = -\dfrac{1}{2}$

     $2x - y = 10$

       $y = 2x - 10$

       $m = 2$

    Since $m_1 \cdot m_2 = -1$, the lines are perpendicular.

12. Let $A = (-6, 5)$, $B = (1, 7)$, and $C = (16, 10)$.
The points would be collinear if
$$m_{\overline{AB}} = m_{\overline{BC}}$$
$$m_{\overline{AB}} = \frac{7-5}{1-(-6)} = \frac{2}{7}$$
$$m_{\overline{BC}} = \frac{10-7}{16-1} = \frac{3}{15} = \frac{1}{5}$$
The points are not collinear.

13. Let $A = (-2, 3)$, $B = (x, 6)$, and $C = (8, 8)$. If $m_{\overline{AB}} = m_{\overline{BC}}$, then A, B, and C are collinear.
$$m_{\overline{AB}} = \frac{6-3}{x-(-2)} = \frac{3}{x+2}$$
$$m_{\overline{BC}} = \frac{8-x}{8-x} = \frac{2}{8-x}$$
$$\frac{3}{x+2} = \frac{2}{8-x}$$
$$3(8-x) = 2(x+2)$$
$$24 - 3x = 2x + 4$$
$$20 = 5x$$
$$x = 4$$

14. Intercepts are (7, 0) and (0, 3).

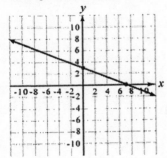

15. $4x - 3y = 9$
$$-3y = -4x + 9$$
$$y = \frac{4}{3}x - 3$$

First, plot the point (0, –3). Then locate a second point for which an increase of 4 in y corresponds to an increase of 3 in x.

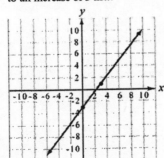

16. $y + 2 = \frac{-2}{3}(x - 1)$

$y - (-2) = \frac{-2}{3}(x-1)$ is a line which contains

$(1, -2)$ and has slope $m = \frac{-2}{3}$. First plot

$(1, -2)$. Then from that point draw a line which

has $m = \frac{-2}{3}$.

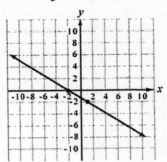

17. a. $m = \frac{6-3}{-3-2} = \frac{3}{-5}$ or $\frac{-3}{5}$
$$y - 3 = \frac{-3}{5}(x-2)$$
$$5(y-3) = -3(x-2)$$
$$5y - 15 = -3x + 6$$
$$3x + 5y = 21$$

b. $m = \frac{-9-(-3)}{8-6} = \frac{-6}{2} = -3$
$$y - (-1) = -3(x-(-2))$$
$$y + 1 = -3(x+2)$$
$$y + 1 = -3x - 6$$
$$3x + y = -7$$

c. $x + 2y = 4$
$$y = \frac{-1}{2}x + 2$$

Since $m_1 = \frac{-1}{2}$, the desired line has $m_2 = 2$.
$$y - (-2) = 2(x-3)$$
$$y + 2 = 2x + 6$$
$$-2x + y = -8$$

d. A line parallel to the x-axis has the form
$y = b$. $\therefore y = 5$

18. Let $A = (-2, -3)$, $B = (4, 5)$, and $C = (-4, 1)$.
$$m_{\overline{AB}} = \frac{5-(-3)}{4-(-2)} = \frac{8}{6} = \frac{4}{3}$$
$$m_{\overline{BC}} = \frac{1-5}{-4-4} = \frac{-4}{-8} = \frac{1}{2}$$
$$m_{\overline{AC}} = \frac{1-(-3)}{-4-(-2)} = \frac{4}{-2} = -2$$

Because $m_{\overline{AC}} \cdot m_{\overline{BC}} = -1$, $\overline{AC} \perp \overline{BC}$ and $\angle C$ is

a rt. $\angle$.

19. Let $A = (3, 6)$, $B = (-6\ 4)$, and $C = (1, -2)$.

$$AB = \sqrt{(-6-3)^2 + (4-6)^2}$$
$$= \sqrt{(-9)^2 + (-2)^2} = \sqrt{81 + 4} = \sqrt{85}$$
$$BC = \sqrt{(1-(-6))^2 + (-2-4)^2}$$
$$= \sqrt{(-9)^2 + (-2)^2} = \sqrt{49 + 36} = \sqrt{85}$$
$$AC = \sqrt{(1-3)^2 + (-2-6)^2}$$
$$= \sqrt{(-2)^2 + (-8)^2} = \sqrt{4 + 64}$$
$$= \sqrt{68} = 2\sqrt{17}$$

Because $AB = BC$, the triangle is isosceles.

20. $R = (-5, -3)$, $S = (1, -11)$, $T = (7, -6)$, and $V = (1, 2)$.

$$m_{\overline{RS}} = \frac{-11 - (-3)}{1 - (-5)} = \frac{-8}{6} = \frac{-4}{3}$$
$$m_{\overline{ST}} = \frac{-6 - (-11)}{7 - 1} = \frac{5}{6}$$
$$m_{\overline{TV}} = \frac{2 - (-6)}{1 - 7} = \frac{8}{-6} = \frac{-4}{3}$$
$$m_{\overline{RV}} = \frac{2 - (-3)}{1 - (-5)} = \frac{5}{6}$$

$\therefore \overline{RS} \parallel \overline{VT}$ and $\overline{RV} \parallel \overline{ST}$ and $RSTV$ is a parallelogram.

21. Solution by graphing:
$$4x - 3y = -3$$
$$y = \frac{4}{3}x + 1$$
$$x + 2y = 13$$
$$y = -\frac{1}{2}x + \frac{13}{2}$$

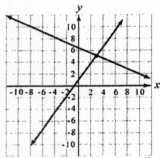

The graphs (lines) intersect at $(3, 5)$.

22. Solution by graphing:
$$y = x + 3$$
$$y = 4x$$

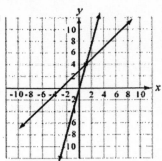

The graphs (lines) intersect at $(1, 4)$.

23. Solution by Substitution:
$$4x - 3y = -3$$
$$x + 2y = 13$$
$$x = 13 - 2y$$
$$4(13 - 2y) - 3y = -3$$
$$52 - 8y - 3y = -3$$
$$-11y = -55$$
$$y = 5$$
$$x = 13 - 2(5) = 3$$

$(3, 5)$

24. Solution by Substitution:
$$4x = x + 3$$
$$3x = 3$$
$$x = 1$$
$$y = 4(1) = 4$$

$(1, 4)$

25.

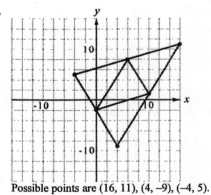

Possible points are $(16, 11)$, $(4, -9)$, $(-4, 5)$.

26. a. $D = M_{\overline{AC}} = (7,2)$

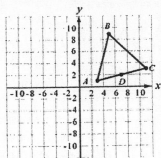

The length of $\overline{BD}$ is

$$\sqrt{(7-5)^2 + (2-9)^2} = \sqrt{2^2 + (-7)^2}$$
$$= \sqrt{4+49} = \sqrt{53}$$

b. $m_{\overline{AC}} = \frac{3-1}{11-3} = \frac{2}{8} = \frac{1}{4}$.

Then the slope of the altitude to $\overline{AC}$ is -4.

c. Since $m_{\overline{AC}} = \frac{1}{4}$, the slope of any line parallel

to $\overline{AC}$ is also $\frac{1}{4}$.

27. $A = (-a, 0)$
$B = (0, b)$
$C = (a, 0)$

28. $D = (0, 0)$
$E = (a, 0)$
$F = (a, 2a)$
$G = (0, 2a)$

29. $R = (0, 0)$
$U = (0, a)$
$T = (a, a+b)$

30. $M = (0, 0)$
$N = (a, 0)$
$Q = (a+b, c)$
$P = (b, c)$

31. a. The midpoint of $\overline{AB}$ is
$M_{\overline{AB}} = (a+c, b+d)$. Then

$$CM = \sqrt{(a+c-0)^2 + (b+d-2e)^2}$$
$$= \sqrt{(a+c)^2 + (b+d-2e)^2}$$

b. $m_{\overline{AC}} = \frac{2e-2b}{0-2a} = \frac{e-b}{-a}$ or $\frac{b-e}{a}$

Then the slope of the altitude to $\overline{AC}$ is

$-\frac{a}{b-e}$ or $\frac{b-e}{a}$.

c. The altitude from B to $\overline{AC}$ contains the point

$(2c, 2d)$ and has slope $m = \frac{a}{e-b}$ from part

(b). $y - 2d = \frac{a}{e-b}(x-2c)$.

32. See Section 9.4, #18.

33. If the diagonals of a rectangle are perpendicular, then the rectangle is a square.

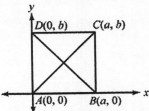

Proof: Let rect. $ABCD$ have vertices as shown.
If $\overline{DB} \perp \overline{AC}$, then $m_{\overline{DB}} \cdot m_{\overline{AC}} = -1$. But

$m_{\overline{DB}} = \frac{0-b}{a-0} = \frac{-b}{a}$ and $m_{\overline{AC}} = \frac{b-0}{a-0} = \frac{b}{a}$. Then

$$\frac{-b}{a} \cdot \frac{b}{a} = -1$$
$$\frac{-b^2}{a^2} = -1$$
$$-b^2 = -a^2$$
$$b^2 = a^2$$

Since a and b are both positive, $a = b$. Then
$AB = a - 0 = a$ and $AD = b - 0 = b$. Since $a = b$,
$AB = AD$. If $AB = AD$, then $ABCD$ is a square.

34. If the diagonals of a trapezoid are equal in length, then the trapezoid is an isosceles trapezoid.

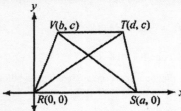

Proof: Let trap. *RSTV* have vertices as shown. If $RT = VS$, then

$$\sqrt{(d-0)^2 + (c-0)^2} = \sqrt{(a-b)^2 + (0-c)^2}$$
$$\sqrt{d^2 + c^2} = \sqrt{(a-b)^2 + (-c)^2}$$
$$\sqrt{d^2 + c^2} = \sqrt{(a-b)^2 + c^2}$$

Squaring, $d^2 + c^2 = (a-b)^2 + c^2$
$$d^2 = (a-b)^2$$
$$d = (a-b$$

Comparing the lengths of $\overline{RV}$ and $\overline{ST}$,

$$RV = \sqrt{(b-0)^2 + (c-0)^2} = \sqrt{b^2 + c^2}$$
$$ST = \sqrt{(a-d)^2 + (0-c)^2} = \sqrt{(a-d)^2 + c^2}$$

Now *RV* would equal *ST* if

$$b^2 + c^2 = (a-d)^2 + c^2$$
$$b^2 = (a-d)^2$$
$$b = a-d$$

Because $d = a - b$ leads to $b = a - d$, so $RV = ST$. Then *RSTV* is isosceles.

35. If two medians of a triangle are equal in length, then the triangle is isosceles.

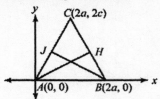

Proof: Let $\triangle ABC$ has vertices as shown, so that the midpoint of $\overline{AC}$ is $J = (b, c)$ and the midpoint of $\overline{BC}$ is $H = (a+b, c)$. If $AH = BJ$, then

$$\sqrt{(a+b-0)^2 + (c-0)^2} = \sqrt{(2a-b)^2 + (0-c)^2}$$
$$\sqrt{(a+b)^2 + c^2} = \sqrt{(2a-b)^2 + c^2}$$

Squaring, $(a+b)^2 + c^2 = (2a-b)^2 + c^2$
$$(a+b)^2 = (2a-b)^2$$

Taking the principal of square roots,
$$a + b = 2a - b$$
$$2b = a$$

Then point $C = (a, 2c)$.

Now $AC = \sqrt{(a-0)^2 + (2c-0)^2}$
$$= \sqrt{a^2 + (2c)^2} = \sqrt{a^2 + 4c^2}$$

and $BC = \sqrt{(2a-a)^2 + (0-2c)^2}$
$$= \sqrt{a^2 + (-2c)^2} = \sqrt{a^2 + 4c^2}$$

Because $AC = BC$, the triangle is isosceles.

36. The segments joining the midpoints of consecutive sides of an isosceles trapezoid form a rhombus.

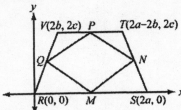

Proof: Let trapezoid *RSTV* have vertices as shown so that $RV = TS$. Where $M, N, P,$ and Q are the midpoints of the sides,

$$M = \left(\frac{0+2a}{2}, \frac{0+0}{2}\right) = (a, 0)$$

$$N = \left(\frac{(2a-2b)+2a}{2}, \frac{0+2c}{2}\right) = (2a-b, c)$$

$$P = \left(\frac{(2a-2b)+2b}{2}, \frac{2c+2c}{2}\right) = (a, 2c)$$

$$Q = \left(\frac{0+2b}{2}, \frac{0+2c}{2}\right) = (b, c)$$

By an earlier theorem, *MNPQ* is a parallelogram. We must show that two adjacent sides of parallelogram *MNPQ* are equal in length.

$$QM = \sqrt{(a-b)^2 + (0-c)^2}$$
$$= \sqrt{(a-b)^2 + (-c)^2} = \sqrt{(a-b)^2 + c^2}$$
$$MN = \sqrt{(2a-b-a)^2 + (c-0)^2} = \sqrt{(a-b)^2 + c^2}$$

Now $QM = MN$, so *MNPQ* is a rhombus.

CHAPTER TEST

1. a. $(5, -3)$

    b. $(0, -4)$

2.

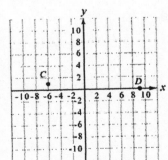

3. $d = \sqrt{(-6-0)^2 + (1-9)^2} = \sqrt{(-6)^2 + (-8)^2}$
$= \sqrt{36 + 64} = \sqrt{100} = 10$

4. $M = \left(\frac{x_1 + x_2}{2}, \frac{y_1 + y_2}{2}\right) = \left(\frac{-6+0}{2}, \frac{1+9}{2}\right)$
    $= (-3, 5)$

5.

x	0	3	0	9
y	4	2	4	-2

6. Graph of $2x + 3y = 12$,

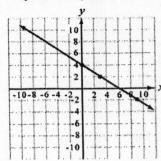

7. a. $m = \frac{-6-3}{2-(-1)} = \frac{-9}{3} = -3$

    b. $m = \frac{d-b}{c-a}$

8. a. $\frac{2}{3}$

    b. $-\frac{3}{2}$

9. $AB = \sqrt{(0-0)^2 + (a-0)^2} = \sqrt{a^2} = a$
    $BC = \sqrt{(a+b-a)^2 + (c-0)^2} = \sqrt{b^2 + c^2}$
    $CD = \sqrt{(b-(a+b))^2 + (c-c)^2} = \sqrt{(-a)^2} = a$
    $AD = \sqrt{(b-0)^2 + (c-0)^2} = \sqrt{b^2 + c^2}$
Since $\overline{AB} = \overline{CD}$ and $\overline{BC} = \overline{AD}$, *ABCD* is a parallelogram.

10. $AB = \sqrt{(0-0)^2 + (a-0)^2} = \sqrt{a^2} = a$
    $AD = \sqrt{(b-0)^2 + (c-0)^2} = \sqrt{b^2 + c^2}$
If $\overline{AB} = \overline{AD}$, then $a = \sqrt{b^2 + c^2}$ or $a^2 = b^2 + c^2$.

11. a. isosceles triangle

    b. trapezoid

12. a. Slope Formula

    b. Distance Formula

13. Let $D = (0, 0)$ and $E = (2a, 0)$, then $F = (a, b)$.

14. (b)

15. Since $RSTV$ is a parallelogram, then $m_{\overline{VR}} = m_{\overline{TS}}$.

$$m_{\overline{VR}} = \frac{v-0}{0-r} = -\frac{v}{r}$$
$$m_{\overline{TS}} = \frac{v-0}{t-s} = \frac{v}{t-s}$$

So $-\frac{v}{r} = \frac{v}{t-s}$

$$rv = -v(t-s)$$
$$r = -t+s$$

or $s = r+t$ or $t = -r+s$.

16. a. $m = \dfrac{6-4}{2-0} = \dfrac{2}{2} = 1$

$$y-4 = 1(x-0)$$
$$y-4 = x$$
$$y = x+4$$

b. $m = \dfrac{3}{4}$, $b = -3$

$$y = \frac{3}{4}x - 3$$

17. The slope of the perpendicular line $m = c$.

$$y-b = c(x-a)$$
$$y-b = cx - ac$$
$$y = cx + (b-ac)$$

18. $x + 2y = 6$

$$x = -2y + 6$$

Use substitution,

$$2x - y = 7$$
$$2(-2y+6) - y = 7$$
$$-4y + 12 - y = 7$$
$$-5y = -5$$
$$y = 1$$
$$x = -2(1) + 6 = 4$$

(4, 1)

19. $5x - 2y = -13$ Multiply by 5
    $3x + 5y = 17$ Multiply by 2

$$25x - 10y = -65$$
$$\underline{6x + 10y = 34}$$
$$31x = -31$$
$$x = -1$$

$$3(-1) + 5y = 17$$
$$-3 + 5y = 17$$
$$5y = 20$$
$$y = 4$$

(–1, 4)

20. $M = (a+b, c)$ and $N = (a, 0)$.

Then $m_{\overline{AC}} = \dfrac{2c-0}{2b-0} = \dfrac{c}{b}$.

Also $m_{\overline{MN}} = \dfrac{c-0}{a+b-a} = \dfrac{c}{b}$.

With $m_{\overline{AC}} = m_{\overline{MN}}$, it follows that $\overline{AC} \parallel \overline{MN}$.

Chapter 10: Introduction to Trigonometry

SECTION 10.1: The Sine Ratio and Applications

1. $\sin\alpha = \dfrac{\text{opposite}}{\text{hypotenuse}} = \dfrac{5}{13}$

$\sin\beta = \dfrac{12}{13}$

5.
$$a^2 + b^2 = c^2$$
$$\left(\sqrt{2}\right)^2 + \left(\sqrt{3}\right)^2 = c^2$$
$$2 + 3 = c^2$$
$$5 = c^2$$
$$\sqrt{5} = c$$

$\sin\alpha = \dfrac{\sqrt{3}}{\sqrt{5}} = \dfrac{\sqrt{3}}{\sqrt{5}} \cdot \dfrac{\sqrt{5}}{\sqrt{5}} = \dfrac{\sqrt{15}}{5}$

$\sin\beta = \dfrac{\sqrt{2}}{\sqrt{5}} = \dfrac{\sqrt{2}}{\sqrt{5}} \cdot \dfrac{\sqrt{5}}{\sqrt{5}} = \dfrac{\sqrt{10}}{5}$

9. $\sin 17° = 0.2924$

13. $\sin 72° = 0.9511$

17. $\sin 43° = \dfrac{a}{16}$

$a = 16\sin 43°$
$a \approx 16(0.6820)$
$a \approx 10.9 \text{ ft}$

$\sin 47° = \dfrac{b}{16}$

$b = 16\sin 47°$
$b \approx 16(0.7314)$
$b \approx 11.7 \text{ ft}$

21. $\sin\alpha = \dfrac{12}{25} = 0.4800$

$\alpha \approx 29°$
$\beta \approx 90° - 29°$ or $\beta \approx 61°$

25. $\sin\alpha = \dfrac{x}{3x} = \dfrac{1}{3} \approx 0.3333$

$\alpha \approx 19°$
$\beta \approx 90° - 19°$ or $\beta \approx 71°$

29. Let d represent the distance between Danny and the balloon.

$\sin 75° = \dfrac{100}{d}$

$d \cdot \sin 75° = 100$

$d = \dfrac{100}{\sin 75°}$

$d \approx \dfrac{100}{0.9659}$

$d \approx 103.5 \text{ ft}$

33. $\sin\alpha = \dfrac{4}{10}$

$\sin\alpha = 0.4000$

$\alpha \approx 24°$

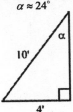

37. $\sin\theta = \dfrac{10}{13} \approx 0.7692$

$\theta \approx 50°$

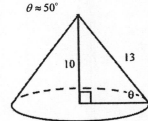

SECTION 10.2: The Cosine Ratio and Applications

1. $\cos\alpha = \dfrac{\text{adjacent}}{\text{hypotenuse}} = \dfrac{12}{13}$

$\cos\beta = \dfrac{5}{13}$

5.
$$a^2 + b^2 = c^2$$
$$\left(\sqrt{3}\right)^2 + \left(\sqrt{2}\right)^2 = c^2$$
$$3 + 2 = c^2$$
$$5 = c^2$$
$$\sqrt{5} = c$$

$\cos\alpha = \dfrac{\sqrt{2}}{\sqrt{5}} = \dfrac{\sqrt{2}}{\sqrt{5}} \cdot \dfrac{\sqrt{5}}{\sqrt{5}} = \dfrac{\sqrt{10}}{5}$

$\cos\beta = \dfrac{\sqrt{3}}{\sqrt{5}} = \dfrac{\sqrt{3}}{\sqrt{5}} \cdot \dfrac{\sqrt{5}}{\sqrt{5}} = \dfrac{\sqrt{15}}{5}$

9. $\cos 23° \approx 0.9205$

13. $\cos 90° = 0$

17. $\cos 32° = \dfrac{a}{100}$

$a = 100\cos 32°$

$a \approx 100(0.8480)$

$a \approx 84.8 \text{ ft}$

$\sin 32° = \dfrac{b}{100}$

$b = 100\sin 32°$

$b \approx 100(0.5299)$

$b \approx 53.0 \text{ ft}$

21. $\cos 51° = \dfrac{12}{c}$

$c = \dfrac{12}{\cos 51°}$

$c \approx \dfrac{12}{0.6293}$

$c \approx 19.1 \text{ ft}$

$\sin 51° = \dfrac{d}{c}$

$d = c \cdot \sin 51°$

$d \approx (19.1)(0.7771)$

$d \approx 14.8 \text{ ft}$

25. $\qquad a^2 + b^2 = c^2$

$\left(\sqrt{3}\right)^2 + \left(\sqrt{2}\right)^2 = c^2$

$3 + 2 = c^2$

$5 = c^2$

$\sqrt{5} = c$

$\cos \alpha = \dfrac{\sqrt{2}}{\sqrt{5}} = \sqrt{\dfrac{2}{5}} = \sqrt{0.4} \approx 0.6325$

$\alpha \approx 51°$

$\cos \beta \approx 90° - 51°$

$\beta \approx 39°$

29. $\cos \theta = \dfrac{10}{12} \approx 0.8333$

$\theta \approx 34°$

33. Let c represent the measure of the central angle of the regular pentagon. Then $c = \dfrac{360°}{5} = 72°$.

Because the apothem shown bisects the central angle,

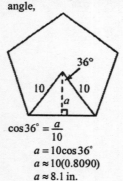

$\cos 36° = \dfrac{a}{10}$

$a = 10\cos 36°$

$a \approx 10(0.8090)$

$a \approx 8.1 \text{ in.}$

37. Let d represent the length of the diagonal of the base and x the length of an edge of the cube. By the Pythagorean Theorem,

$d^2 = x^2 + x^2$

$d^2 = 2x^2$.

Let D represent the Pythagorean Theorem again,

$x^2 + d^2 = D^2$

$x^2 + 2x^2 = D^2$

so that

$D^2 = 3x^2$

$D = \sqrt{3x^2}$

$D = x\sqrt{3}$

Now, $\cos \alpha = \dfrac{\text{adjacent}}{\text{hypotenuse}} = \dfrac{x}{x\sqrt{3}} = \dfrac{1}{\sqrt{3}}$

$\cos \alpha \approx 0.5774$

$\alpha \approx 55°$

41.

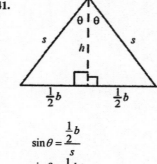

$\sin \theta = \dfrac{\frac{1}{2}b}{s}$

$s \cdot \sin \theta = \dfrac{1}{2}b$

$b = 2s \cdot \sin \theta$

$\cos \theta = \dfrac{h}{s}$

$h = s \cdot \cos \theta$

$A_{\vartriangle} = \dfrac{1}{2}bh$

$A_{\vartriangle} = \dfrac{1}{2}(2s \cdot \sin \theta)(s \cdot \cos \theta)$

$A_{\vartriangle} = s^2 \cdot \sin \theta \cdot \cos \theta$

SECTION 10.3: The Tangent Ratio and Other Ratios

1. $\tan \alpha = \dfrac{\text{opposite}}{\text{adjacent}} = \dfrac{3}{4}$

$\tan \beta = \dfrac{4}{3}$

5. Using the Pythagorean Triple, (5, 12, 13), $b = 12$.

$\sin\alpha = \dfrac{\text{opposite}}{\text{hypotenuse}} = \dfrac{5}{13}$

$\cos\alpha = \dfrac{\text{adjacent}}{\text{hypotenuse}} = \dfrac{12}{13}$

$\tan\alpha = \dfrac{\text{opposite}}{\text{adjacent}} = \dfrac{5}{12}$

$\cot\alpha = \dfrac{\text{adjacent}}{\text{opposite}} = \dfrac{12}{5}$

$\sec\alpha = \dfrac{\text{hypotenuse}}{\text{adjacent}} = \dfrac{13}{12}$

$\csc\alpha = \dfrac{\text{hypotenuse}}{\text{opposite}} = \dfrac{13}{5}$

9. $a^2 + b^2 = c^2$

$x^2 + b^2 = \left(\sqrt{x^2+1}\right)^2$

$x^2 + b^2 = x^2 + 1$

$b^2 = 1$

$b = 1$

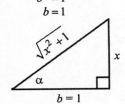

$\sin\alpha = \dfrac{x}{\sqrt{x^2+1}}$

$= \dfrac{x}{\sqrt{x^2+1}} \cdot \dfrac{\sqrt{x^2+1}}{\sqrt{x^2+1}}$

$= \dfrac{x\sqrt{x^2+1}}{x^2+1}$

$\cos\alpha = \dfrac{1}{\sqrt{x^2+1}}$

$= \dfrac{1}{\sqrt{x^2+1}} \cdot \dfrac{\sqrt{x^2+1}}{\sqrt{x^2+1}}$

$= \dfrac{\sqrt{x^2+1}}{x^2+1}$

$\tan\alpha = \dfrac{x}{1}$

$\cot\alpha = \dfrac{1}{x}$

$\sec\alpha = \dfrac{\sqrt{x^2+1}}{1} = \sqrt{x^2+1}$

$\csc\alpha = \dfrac{\sqrt{x^2+1}}{x}$

13. $\tan 57° = 1.5399$

17. $\cos 58° = \dfrac{y}{10}$

$y = 10\cos 58°$

$y \approx 10(0.5299)$

$y \approx 5.3$

$\sin 58° = \dfrac{z}{10}$

$z = 10\sin 58°$

$z \approx 10(0.8480)$

$z \approx 8.5$

21. $\tan\alpha = \dfrac{3}{4} = 0.7500$

$\alpha \approx 37°$

$\beta \approx 90° - 37°$ or $\beta \approx 53°$

25. $\tan\alpha = \dfrac{\sqrt{5}}{4} \approx \dfrac{2.2361}{4} \approx 0.5590$

$\alpha \approx 29°$

$\beta \approx 90° - 29°$ or $\beta \approx 61°$

29. $\csc 30° = \dfrac{1}{\sin 30°} = \dfrac{1}{0.5} = 2.0000$ (exact)

33. a. $\sin\alpha = \dfrac{a}{c}$ and $\cos\alpha = \dfrac{b}{c}$

Then $\dfrac{\sin\alpha}{\cos\alpha} = \sin\alpha \div \cos\alpha = \dfrac{a}{c} \div \dfrac{b}{c}$.

In turn, $\dfrac{\sin\alpha}{\cos\alpha} = \dfrac{a}{c} \cdot \dfrac{c}{b} = \dfrac{a}{b} = \tan\alpha$.

b. $\tan 23° \approx 0.4245$

$\dfrac{\sin 23°}{\cos 23°} \approx \dfrac{0.39073}{0.92050} \approx 0.4245 \approx \tan 23°$

37. $\sin 5° = \dfrac{120}{x}$

$x \cdot \sin 5° = 120$

$x = \dfrac{120}{\sin 5°}$

$x \approx \dfrac{120}{0.0872} \approx 1376.8$ ft

41.

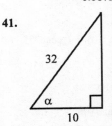

$\cos\alpha = \dfrac{10}{32}$

$\cos\alpha = 0.3125$

$\alpha \approx 72°$

45.

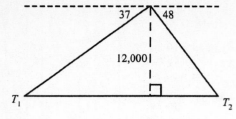

$$\tan 37 = \frac{12,000}{x}$$
$$x \cdot \tan 37 = 12,000$$
$$x = \frac{12,000}{\tan 37}$$
$$x \approx 15,924.5$$

$$\tan 48 = \frac{12,000}{y}$$
$$y \cdot \tan 48 = 12,000$$
$$y = \frac{12,000}{\tan 48}$$
$$y \approx 10,804.8$$

Distance $\approx 15,924.5 + 10,804.8$
$$\approx 26,729.3 \approx 26,730 \text{ feet}$$

SECTION 10.4: Applications to Acute Triangles

1. a. $A = \frac{1}{2} ab \sin \gamma$

$$A = \frac{1}{2} \cdot 5 \cdot 6 \cdot \sin 78°$$

b. $\alpha + \beta + \gamma = 180°$
$$36° + 88° + \gamma = 180°$$
$$124 + \gamma = 180$$
$$\gamma = 56°$$

$$A = \frac{1}{2} ab \sin \gamma$$
$$A = \frac{1}{2} \cdot 5 \cdot 7 \cdot \sin 56°$$

5. a. $c^2 = a^2 + b^2 - 2ab \cos \gamma$
$$c^2 = (5.2)^2 + (7.9)^2 - 2(5.2)(7.9) \cos 83°$$

b. $a^2 = b^2 + c^2 - 2bc \cos \alpha$
$$6^2 = 9^2 + 10^2 - 2(9)(10) \cos \alpha$$

9. a. (3, 4, 5) is a Pythagorean Triple. γ lies opposite the longest side and must be a right angle.

b. 90°

13. With measures of angles shown, the third angle measures 70°. Then the triangle is isosceles, with sides as shown.

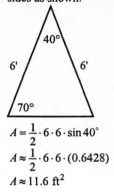

$$A = \frac{1}{2} \cdot 6 \cdot 6 \cdot \sin 40°$$
$$A \approx \frac{1}{2} \cdot 6 \cdot 6 \cdot (0.6428)$$
$$A \approx 11.6 \text{ ft}^2$$

17. $\dfrac{\sin \alpha}{a} = \dfrac{\sin \beta}{b}$
$$\frac{\sin 60°}{x} = \frac{\sin 70°}{12}$$
$$\frac{0.8660}{x} = \frac{0.9397}{12}$$
$$x(0.9397) = 12(0.8660)$$
$$x \approx 11.1 \text{ in.}$$

21. $\dfrac{\sin \gamma}{10} = \dfrac{\sin 80°}{12}$
$$\frac{\sin \gamma}{10} = \frac{0.9397}{12}$$
$$12(\sin \gamma) = 10(0.9848)$$
$$\sin \gamma \approx 0.8207$$
$$\gamma \approx 55°$$

25. $x^2 = 8^2 + 12^2 - 2 \cdot 8 \cdot 12 \cos 60°$
$$x^2 = 64 + 144 - 192 \cos 60°$$
$$x^2 = 208 - 192(0.5)$$
$$x^2 = 112$$
$$x = \sqrt{112} \approx 10.6$$

29. a. $x^2 = 150^2 + 180^2 - 2(150)(180) \cos 80°$
$$x^2 = 22,500 + 32,400 - 54,000(0.1736)$$
$$x^2 = 44,525.6$$
$$x = \sqrt{44,525.6}$$
$$x \approx 213.4 \text{ feet}$$

b. $A = \frac{1}{2}(150)(180) \sin 80°$
$$A \approx \frac{1}{2}(150)(180)(0.9848)$$
$$A \approx 13,294.9 \text{ ft}^2$$

33. The third angle measures 85°.
Using the Law of Sines,

$$\frac{\sin 85°}{x} = \frac{\sin 30°}{8}$$

$$\frac{0.9962}{x} = \frac{0.5}{8}$$

$$0.5x = 8(0.9962)$$

$$x \approx 15.9 \text{ ft}$$

37. $a^2 = 27^2 + 27^2 - 2(27)(27)\cos 30°$

$a^2 = 729 + 729 - 2(729) \cdot \frac{\sqrt{3}}{2}$

$a^2 \approx 195.33$

$a \approx 13.97 \approx 14.0 \text{ feet}$

41. $A = (6.3)(8.9)\sin 67.5°$

$A = 51.8 \text{ cm}^2$

CHAPTER REVIEW

1. sine;

$$\sin 40° = \frac{a}{16}$$

$$a = 16 \sin 40°$$

$$a \approx 10.3 \text{ in.}$$

2. sine;

$$\sin 70° = \frac{d}{8}$$

$$d = 8 \sin 70°$$

$$d \approx 7.5 \text{ ft}$$

3. cosine;

$$\cos 80° = \frac{4}{c}$$

$$c = \frac{4}{\cos 80°}$$

$$c \approx \frac{4}{0.1736}$$

$$c \approx 23 \text{ in.}$$

4. sine;

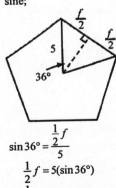

$$\sin 36° = \frac{\frac{1}{2}f}{5}$$

$$\frac{1}{2}f = 5(\sin 36°)$$

$$\frac{1}{2}f \approx 5(0.5878)$$

$$\frac{1}{2}f \approx 2.939$$

$$f \approx 5.9 \text{ ft}$$

5. tangent;

$$\tan \alpha = \frac{13}{14}$$

$$\tan \alpha \approx 0.9286$$

$$\alpha \approx 43°$$

6. cosine;

$$\cos \theta = \frac{8}{15}$$

$$\cos \theta \approx 0.5333$$

$$\theta \approx 58°$$

7. sine;

$$\sin \alpha = \frac{9}{12}$$

$$\sin \alpha \approx 0.7500$$

$$\alpha \approx 49°$$

8. tangent;

$$\tan \beta = \frac{7}{24}$$

$$\tan \beta \approx 0.2917$$

$$\beta \approx 16°$$

9. Law of Sines

$$\frac{\sin 57°}{x} = \frac{\sin 49°}{8}$$

$$x(\sin 49°) = 8(\sin 57°)$$

$$x(0.7547) = 8(0.8387)$$

$$x \approx 8.9 \text{ units}$$

10. Law of Cosines

$$15^2 = 14^2 + 16^2 - 2(14)(16)\cos \alpha$$

$$225 = 196 + 256 - 448 \cos \alpha$$

$$448 \cos \alpha = 227$$

$$\cos \alpha = \frac{227}{448}$$

$$\cos \alpha \approx 0.5067$$

$$\alpha \approx 60°$$

11. The third angle measures 80°.
Law of Sines

$$\frac{\sin 40°}{y} = \frac{\sin 80°}{20}$$

$$y(\sin 80°) = 20(\sin 40°)$$

$$y(0.9848) = 20(0.6428)$$

$$y \approx 13.1$$

12. Law of Cosines

$$w^2 = 14^2 + 21^2 - 2(14)(21)\cos 60°$$

$$w^2 = 196 + 441 - 294$$

$$w^2 = 343$$

$$w = \sqrt{343}$$

$$w \approx 18.5$$

13. The remaining angles of the triangle measure 47°
    and 74°.

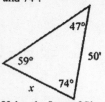

Using the Law of Sines,

$$\frac{\sin 47°}{x} = \frac{\sin 59°}{50}$$

$$\frac{0.7314}{x} = \frac{0.8572}{50}$$

$$x(0.8572) = 50(0.7314)$$

$$x \approx 42.7 \text{ feet}$$

14. Let d be the length of the shorter diagonal.

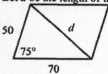

Law of Cosines

$$d^2 = 50^2 + 70^2 - 2(50)(70)\cos 75°$$

$$d^2 = 2500 + 4900 - 7000(0.2588)$$

$$d^2 = 2500 + 4900 - 1811.6$$

$$d^2 = 5588.4$$

$$d = \sqrt{5588.4}$$

$$d \approx 74.8 \text{ cm}$$

15. Law of Cosines

$$6^2 = 6^2 + 11^2 - 2(6)(11)\cos\alpha$$

$$36 = 366 + 121 - 132\cos\alpha$$

$$132\cos\alpha = 121$$

$$\cos\alpha = \frac{121}{132}$$

$$\cos\alpha \approx 0.9167$$

$$\alpha \approx 23.6°$$

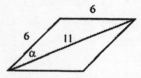

The acute angle of the rhombus measures
$2\alpha \approx 47°$.

16.

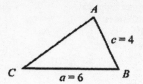

$$A = \frac{1}{2}ac\sin B$$

$$9.7 = \frac{1}{2}(6)(4)\sin B$$

$$9.7 = 12\sin B$$

$$\sin B = \frac{9.7}{12} \approx 0.8083$$

$$B \approx 54°$$

17.

If the acute angle measures 47°, then
$A = \frac{1}{2}ab\sin 47°$ gives one-half the desired area.

$$A = \frac{1}{2} \cdot 6 \cdot 6 \cdot \sin 47°$$

$$A \approx 13.16 \text{ in}^2$$

for $\triangle ABC$. The area of rhombus $ABCD$ is
approximately 26.3 in².

18. If $m\angle R = 45°$,

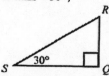

then $m\angle S = 45°$ also. Then $\overline{RT} = \overline{ST}$. Let
$RT = ST = x$. Now, $\tan R = \tan 45° = \frac{x}{x} = 1$.

19. If $m\angle S = 30°$,

then the sides of $\triangle RQS$ can be represented by
$RQ = x$, $RS = 2x$, and $SQ = x \cdot \sqrt{3}$.

$$\sin S = \sin 30° = \frac{x}{2x} = \frac{1}{2}.$$

20. If $m\angle T = 60°$,

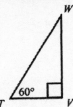

then the sides of $\triangle TVW$ can be represented by $TV = x$, $TW = 2x$, and $VW = x\sqrt{3}$.

$$\sin T = \sin 60° = \frac{x\sqrt{3}}{2x} = \frac{\sqrt{3}}{2}.$$

21. Because alt. int. $\angle$s are $\cong$,

$$\tan 55° = \frac{12}{x}$$
$$x(\tan 55°) = 12$$
$$x = \frac{12}{\tan 55°} \approx \frac{12}{1.4281}$$
$$x \approx 8.4 \text{ ft}$$

22. $\sin 60° = \dfrac{x}{200}$

$$x = 200\sin 60°$$
$$x \approx 200(0.866)$$
$$x \approx 173.2$$

If the rocket rises 173.2 feet per second, then its altitude after 5 seconds will be approximately 866 feet.

23. $\cos \alpha = \dfrac{3}{4}$

$$\cos \alpha = 0.75$$
$$\alpha \approx 41°$$

24. $\sin \alpha = \dfrac{300}{2200}$

$$\sin \alpha = 0.1364$$
$$\alpha \approx 8°$$

25. Let x represent one-half the length of a side of a regular pentagon.

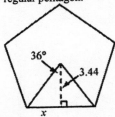

$$\tan 36° = \frac{x}{3.44}$$
$$x = 3.44(\tan 36°)$$
$$x \approx 3.44(0.7265)$$
$$x \approx 2.50$$

Then the length of each side is approximately 5.0 cm.

26.

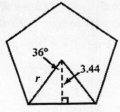

$$\cos 36° = \frac{3.44}{r}$$
$$r = \frac{3.44}{\cos 36°}$$
$$r \approx \frac{3.44}{0.8090}$$
$$r \approx 4.3 \text{ cm}$$

27. The altitude bisects the base where β represents the measure of the base angle,

$$\cos \beta = \frac{15}{40}$$
$$\cos \beta = 0.375$$
$$\beta \approx 68°$$

28. The measure of acute angle α is one-half the desired angle's measure.

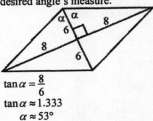

$$\tan \alpha = \frac{8}{6}$$
$$\tan \alpha \approx 1.333$$
$$\alpha \approx 53°$$

29.

$$\tan 23° = \frac{a}{b}$$
$$\tan 23° = \frac{3}{b}$$
$$b(\tan 23°) = 3$$
$$b = \frac{3}{\tan 23°}$$
$$b \approx \frac{3}{0.4245} \approx 7$$

The grade of the hill is 3 to 7 (or 3:7).

30. Let S_1 represent the distance to the nearer ship and S_2 represent the distance to the farther ship.

$$\tan 32° = \frac{2500}{S_2} \quad \text{and} \quad \tan 44° = \frac{2500}{S_1}$$

$$S_2 = \frac{2500}{\tan 32°} \qquad S_1 = \frac{2500}{\tan 44°}$$

$$S_2 \approx 4000.8 \qquad S_1 \approx 2588.8$$

The distance between the ships is
$4000.8 - 2588.8$ or approximately 1412.0 meters.

31. $\sin\theta = \dfrac{7}{25}$

Using $\sin^2\theta + \cos^2\theta = 1$,

$$\left(\frac{7}{25}\right)^2 + \cos^2\theta = 1$$

$$\frac{49}{625} + \cos^2\theta = 1$$

$$\cos^2\theta = 1 - \frac{49}{625}$$

$$\cos^2\theta = \frac{576}{625}$$

$$\cos\theta = \sqrt{\frac{576}{625}} = \frac{\sqrt{576}}{\sqrt{625}} = \frac{24}{25}$$

Because $\sec\theta = \dfrac{1}{\cos\theta}$,

$$\sec\theta = \frac{1}{\frac{24}{25}} = \frac{25}{24}.$$

32. $\tan\theta = \dfrac{11}{60}$

Using $\tan^2\theta + 1 = \sec^2\theta$,

$$\left(\frac{11}{60}\right)^2 + 1 = \sec^2\theta$$

$$\frac{121}{3600} + 1 = \sec^2\theta$$

$$\frac{3721}{3600} = \sec^2\theta$$

$$\sec\theta = \sqrt{\frac{3721}{3600}} = \frac{\sqrt{3721}}{\sqrt{3600}} = \frac{61}{60}$$

Because $\cot\theta = \dfrac{1}{\tan\theta}$,

$$\cot\theta = \frac{1}{\frac{11}{60}} = \frac{60}{11}.$$

33. $\cot\theta = \dfrac{21}{20}$

Using $\cot^2\theta + 1 = \csc^2\theta$,

$$\left(\frac{21}{20}\right)^2 + 1 = \csc^2\theta$$

$$\frac{841}{400} = \csc^2\theta$$

$$\csc\theta = \sqrt{\frac{841}{400}} = \frac{\sqrt{841}}{\sqrt{400}} = \frac{29}{20}$$

Because $\sin\theta = \dfrac{1}{\csc\theta}$,

$$\sin\theta = \frac{1}{\frac{29}{20}} = \frac{20}{29}.$$

34.

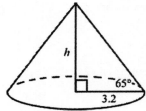

$$\tan 65° = \frac{h}{3.2}$$
$$h = 3.2 \cdot \tan 65$$
$$h \approx 6.9 \text{ feet}$$

$$V = \frac{1}{3}Bh$$
$$V = \frac{1}{3}\pi r^2 h$$
$$V \approx \frac{1}{3}\pi \cdot (3.2)^2 \cdot (6.9)$$
$$V \approx 74.0 \text{ ft}^3$$

CHAPTER TEST

1. a. $\sin\alpha = \dfrac{a}{c}$

　　b. $\tan\beta = \dfrac{b}{a}$

2. a. $\cos\beta = \dfrac{6}{10} = \dfrac{3}{5}$

　　b. $\sin\alpha = \dfrac{6}{10} = \dfrac{3}{5}$

3. a. $\tan 45° = \dfrac{1}{1} = 1$

　　b. $\tan 60° = \dfrac{\sqrt{3}}{2}$

4. a. $\sin 23° \approx 0.3907$

   b. $\cos 79° \approx 0.1908$

5. $\sin\theta = 0.6691$

   $\theta \approx 42°$

6. a. $\tan 26°$

   b. $\cos 47°$

7.

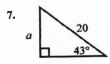

$\sin 43° = \dfrac{a}{20}$

$a = 20\sin 43°$

$a \approx 20(0.6820)$

$a \approx 14$

8.

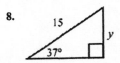

$\sin 37° = \dfrac{y}{15}$

$y = 15\sin 37°$

$y \approx 15(0.6018)$

$y \approx 9$

9.

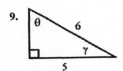

$\sin\theta = \dfrac{5}{6}$

$\sin\theta \approx 0.8333$

$\theta \approx 56°$

10. a. $\cos\beta = \dfrac{a}{c} = \sin\alpha$

   True

   b. True

11. $\sin 67° = \dfrac{x}{100}$

$x = 100\sin 67°$

$x \approx 100(0.9205)$

$x \approx 92$ feet

12. $\sin\theta = \dfrac{2}{12}$

$\sin\theta \approx 0.1667$

$\theta \approx 10°$

13. a. $\csc\alpha = \dfrac{1}{\sin\alpha} = \dfrac{1}{\frac{1}{2}} = 2$

   b. $\sin\alpha = \dfrac{1}{2}$

$\sin\alpha = 0.5$

$\alpha = 30°$

14. a.

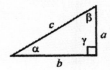

$\cos\beta = \dfrac{a}{c}$

$\sin\alpha = \dfrac{a}{c}$

   b.

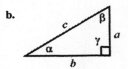

$\cos\beta = \dfrac{a}{c}$

$\sec\beta = \dfrac{1}{\cos\beta}$

$\sec\beta = \dfrac{1}{\frac{a}{c}} = \dfrac{c}{a}$

15. $A = \dfrac{1}{2}bc\sin\alpha$

$A = \dfrac{1}{2}(8)(12)\sin 60°$

$A \approx 48(0.8660)$

$A \approx 42$ cm^2

16.

$\dfrac{\sin\alpha}{a} = \dfrac{\sin\beta}{b} = \dfrac{\sin\gamma}{c}$

17.

$a^2 = b^2 + c^2 - 2bc\cos\alpha$

18. Law of Sines

$$\frac{\sin \alpha}{a} = \frac{\sin \beta}{b}$$

$$\frac{\sin 65°}{10} = \frac{\sin \alpha}{6}$$

$$10 \sin \alpha = 6 \sin 65$$

$$\sin \alpha = \frac{6 \sin 65}{10}$$

$$\sin \alpha \approx 0.5438$$

$$\alpha \approx 33°$$

19. Law of Cosines

$$x^2 = 12^2 + 8^2 - 2(12)(8) \cos 60°$$

$$x^2 = 144 + 64 - 192(0.5)$$

$$x^2 = 112$$

$$x = \sqrt{112}$$

$$x \approx 11$$

Appendix A: Algebra Review

SECTION A.1: Algebraic Expressions

1. Undefined terms, definitions, axioms or postulates, theorems

2. Algebra; geometry

3. **a.** Reflexive

   b. Transitive

   c. Substitution

   d. Symmetric

4. **a.** $2 = 2$

   b. If $2 = x$, then $x = 2$.

   c. If $x = 2$ and $2 = y$, then $x = y$.

   d. If $2 + 5 = 7$ and $7 + y = z$, then $2 + 5 + y = z$.

5. **a.** 12

   b. -2

   c. 2

   d. -12

6. **a.** 8

   b. -8

   c. -22

   d. 1

7. **a.** 35

   b. -35

   c. -35

   d. 35

8. **a.** -84

   b. 84

   c. -84

   d. 84

9. No; Commutative Axiom for Multiplication

10. **a.** Commutative Axiom for Multiplication

    b. Associative Axiom for Addition

    c. Commutative Axiom for Addition

    d. Associative Axiom for Multiplication

11. **a.** 9

    b. -9

    c. 8

    d. -8

12. $(-3) - 7$

13. **a.** -4

    b. -36

    c. 18

    d. $-\dfrac{1}{4}$

14. 5 feet divided by 10 spaces = $\dfrac{1}{2}$ ft per space. Or since 5 feet = 60 inches, 60 inches divided by 10 spaces = 6 inches per space.

15. $-\$60$

16. $25(2) + 30(2) = 50 + 60 = \$110$

17. **a.** $30 + 35 = 65$

    b. $28 - 12 = 16$

    c. $\dfrac{7}{2} + \dfrac{11}{2} = \dfrac{18}{2} = 9$

    d. $8x$

18. **a.** $54 - 24 = 30$

    b. $3(4 + 8) = 12 + 24 = 36$

    c. $7y - 2y = (7 - 2)y = 5y$

    d. $(16 + 8)x = 24x$

19. **a.** $(6 + 4)\pi = 10\pi$

    b. $(8 + 3)\sqrt{2} = 11\sqrt{2}$

    c. $7y - 2y = (7 - 2)y = 5y$

    d. $(9 - 2)\sqrt{3} = 7\sqrt{3}$

20. **a.** $(1 + 2)\pi r^2 = 3\pi r^2$

    b. $(7 + 3)xy = 10xy$

    c. $7x^2y + 3xy^2$

    d. $(1 + 1)x + y = 2x + y$

21. a. $2 + 12 = 14$

b. $5 \cdot 4 = 20$

c. $2 + 3 \cdot 4 = 2 + 12 = 14$

d. $2 + 6^2 = 2 + 36 = 38$

22. a. $9 + 16 = 25$

b. $7^2 = 49$

c. $9 + 6 \div 3 = 9 + 2 = 11$

d. $[9 + 6] \div 3 = 15 \div 3 = 5$

23. a. $\dfrac{6}{-6} = -1$

b. $\dfrac{8 - 6}{6 \cdot 3} = \dfrac{2}{18} = \dfrac{1}{9}$

c. $\dfrac{10 - 18}{9} = \dfrac{-8}{9}$

d. $\dfrac{5 - 12 + (-3)}{4 + 16} = \dfrac{-10}{20} = \dfrac{-1}{2}$

24. a. $8 + 10 + 12 + 15 = 45$

b. $42 + 7 - 12 - 2 = 35$

25. a. $15 - 6 - 5 + 2 = 6$

b. $12x^2 - 15x + 8x - 10 = 12x^2 - 7x - 10$

26. a. $10x^2 - 35x + 6x - 21 = 10x^2 - 39x - 21$

b. $6x^2 - 10xy + 3xy - 5y^2 = 6x^2 - 7xy - 5y^2$

27. $5x + 2y$

28. $2xy + 2yz + 2xz$

29. $10x + 5y$

30. $xy + xz + y^2 + yz$; the total of the areas of the four smaller plots is also $xy + y^2 + xz + yz$.

31. $10x$

32. $9\pi + 48\pi + 9\pi = 66\pi$

SECTION A.2: Formulas and Equations

1. $5x + 8$

2. $-1x + 8$

3. $2x - 2$

4. $4x - 2$

5. $2x + 2 + 3x + 6 = 5x + 8$

6. $6x + 15 - 6x + 2 = 17$

7. $x^2 + 4x + 3x + 12 = x^2 + 7x + 12$

8. $x^2 - 7x - 5x + 35 = x^2 - 12x + 35$

9. $6x^2 - 4x + 5x - 10 = 6x^2 + 11x - 10$

10. $6x^2 + 9x + 14x + 21 = 6x^2 + 23x + 21$

11. $(a + b)(a + b) + (a - b)(a - b)$
$= a^2 + ab + ab + b^2 + a^2 - ab - ab + b^2$
$= 2a^2 + 2b^2$

12. $(x + 2)(x + 2) - (x - 2)(x - 2)$
$= x^2 + 2x + 2x + 4 - \left(x^2 - 2x - 2x + 4\right)$
$= x^2 + 4x + 4 - \left(x^2 - 4x + 4\right)$
$= x^2 + 4x + 4 - x^2 + 4x - 4$
$= 8x$

13. $4 \cdot 3 \cdot 5 = 60$

14. $5^2 + 7^2 = 35 + 49 = 74$

15. $2 \cdot 13 + 2 \cdot 7 = 26 + 14 = 40$

16. $6 \cdot 16 \div 4 = 96 \div 4 = 24$

17. $S = 2 \cdot 6 \cdot 4 + 2 \cdot 4 \cdot 5 + 2 \cdot 6 \cdot 5$
$S = 48 + 40 + 60$
$S = 148$

18. $A = \left(\dfrac{1}{2}\right) 2(6 + 8 + 10)$
$A = 1(24)$
$A = 24$

19. $V = \left(\dfrac{1}{3}\right)\pi(3)^2 \cdot 4$
$V = \dfrac{1}{3} \cdot \pi \cdot 9 \cdot 4$
$V = 12\pi$

20. $S = 4\pi r^2$
$S = 4\pi r(2)^2$
$S = 4\pi \cdot 4$
$S = 16\pi$

21. $2x = 14$
$\qquad x = 7$

22. $3x = -3$
$\qquad x = -1$

23. $\dfrac{y}{-3} = 4$
$\qquad y = -12$

24. $7y = -21$
$\qquad y = -3$

25. $2a + 2 = 26$
$\qquad 2a = 24$
$\qquad\quad a = 12$

26. $\dfrac{3b}{2} = 27$
$\qquad 3b = 54$
$\qquad\quad b = 18$

27. $2x + 2 = 30 - 6x + 12$
$\qquad 2x + 2 = 42 - 6x$
$\qquad\quad 8x = 40$
$\qquad\quad\;\; x = 5$

28. $2x + 2 + 3x + 6 = 22 + 40 - 4x$
$\qquad\quad 5x + 8 = 62 - 4x$
$\qquad\qquad 9x = 54$
$\qquad\qquad\;\; x = 6$

29. Multiply equation by 6 to get
$\qquad 2x - 3x = -30$
$\qquad\qquad -x = -30$
$\qquad\qquad\;\; x = 30$

30. Multiply equation by 12 to get
$\qquad 6x + 4x + 3x = 312$
$\qquad\qquad\quad 13x = 312$
$\qquad\qquad\qquad x = 24$

31. Multiply equation by n to get
$\qquad 360 + 135n = 180n$
$\qquad\qquad\; 360 = 45n$
$\qquad\qquad\quad\; 8 = n$

32. Multiply equation by n to get
$\qquad (n-2)180 = 150n$
$\qquad 180n - 360 = 150n$
$\qquad\qquad\; -360 = -30n$
$\qquad\qquad\qquad 12 = n$

33. $148 = 2 \cdot 5 \cdot w + 2 \cdot w \cdot 6 + 2 \cdot 5 \cdot 6$
$\qquad 148 = 10w + 12w + 60$
$\qquad\; 88 = 22w$
$\qquad\quad 4 = w$

34. $156 = \left(\dfrac{1}{2}\right) \cdot 12 \cdot (b + 11)$
$\qquad 156 = 6(b + 11)$
$\qquad 156 = 6b + 66$
$\qquad\; 90 = 6b$
$\qquad\; 15 = b$

35. $23 = \left(\dfrac{1}{2}\right)(78 - y)$
$\qquad 46 = 78 - y$
$\qquad -32 = -y$
$\qquad\; 32 = y$

36. $\dfrac{-3}{2} = \dfrac{Y - 1}{2 - (-2)}$
$\qquad \dfrac{-3}{2} = \dfrac{Y - 1}{4}$
$\qquad 2Y - 2 = -12$
$\qquad\quad 2Y = -10$
$\qquad\qquad Y = -5$

37. Since $a \cdot c$ is the product of two real numbers, it is a real number, too.
$\therefore a \cdot c = a \cdot c$ (Reflexive Axiom)
But $a = b$ (Hypothesis)
$\therefore a \cdot c = b \cdot c$ (Substitution)

38. Because $a - c$ is the difference of two real numbers, it is a real number, too.
$\therefore a - c = a - c$ (Reflexive Axiom)
But $a = b$ (Hypothesis)
$\therefore a - c = b - c$ (Substitution)

39. Let $a = 2$, $b = 3$, $c = 0$;
$\qquad 2 \cdot 0 = 3 \cdot 0$ but $2 \neq 3$.

40. Let $a = 3$ and $b = -3$;
$\qquad (3)^2 = (-3)^2$ but $3 \neq -3$.

SECTION A.3: Early Definitions and Postulates

1. The length of $\overline{AB}$ is greater than the length of $\overline{CD}$.

2. $e < f;\ f > e$

3. The measure of angle ABC is greater than the measure of angle DEF.

4. $x = 6, x = 9, x = 12$

5. **a.** $p = 4$

    b. $p = 10$

6. Yes

7. $AB > IJ$

8. The measure of angle *JKL* is greater than the measure of angle *ABC*.

9. **a.** False

   b. True

   c. True

   d. False

10. **a.** True

   b. False

   c. True

   d. False

11. The measure of the second angle must be greater than 148° and less than 180°.

12. The length of the second board must be less than 5 feet.

13. **a.** $-12 \leq 20$

   b. $-10 \leq -2$

   c. $18 \geq -30$

   d. $3 \geq -5$

14. **a.** $2 > -1$

   b. $12 < 18$

   c. $-12 > -18$

   d. $2 < 3$

15.

No Change	No Change
No Change	No Change
No Change	CHANGE
No Change	CHANGE

16. $5x \leq 30$
   $x \leq 6$
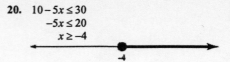

17. $2x \leq 14$
   $x \leq 7$
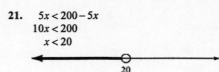

18. $4x > 20$
   $x > 5$
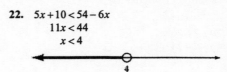

19. $-4x > 20$
   $x < -5$
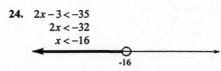

20. $10 - 5x \leq 30$
   $-5x \leq 20$
   $x \geq -4$
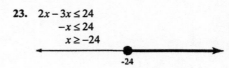

21. $5x < 200 - 5x$
   $10x < 200$
   $x < 20$

22. $5x + 10 < 54 - 6x$
   $11x < 44$
   $x < 4$

23. $2x - 3x \leq 24$
   $-x \leq 24$
   $x \geq -24$

24. $2x - 3 < -35$
   $2x < -32$
   $x < -16$

25. $x^2 + 4x \leq x^2 - 5x - 18$
   $9x \leq -18$
   $x \leq -2$

26. $x^2 + 2x < 2x - x^2 + 2x^2$
   $x^2 + 2x < x^2 + 2x$
   No solution or $\varnothing$.

27. Not true if $c < 0$.

28. Not true if $c = 0$.

29. Not true if $a = -3$ and $b = -2$.

30. Not true if $a = c$.

31. If $a < b$ and $c < d$, then $a + p_1 = b$ and $c + p_2 = d$, where p_1 and p_2 are both positive. Use Addition Property to get
   $a + p_1 + c + p_2 = b + d$
   $a + c + p_1 + p_2 = b + d$
   But since both p_1 and p_2 are both positive, $p_1 + p_2$ must also be positive.
   $\therefore a + c < b + d$.

32. If $a < b$, then $a + p_1 = b$ where p_1 is positive.
Multiply by -1.
$-a - p_1 = -b$
Add p_1 to both sides.
$-a = -b + p_1$
Add c to both sides.
$c + (-a) = c + (-b) + p_1$
$c - a = c - b + p_1$
Since p_1 is positive, $c - a > c - b$.

SECTION A.4: Early Definitions and Postulates

1. a. 3.61

  b. 2.83

  c. -5.39

  d. 0.77

2. a. 4.12

  b. 20

  c. -2.65

  d. 1.26

3. a, c, d, f

4. a, b, c, e

5. a. $\sqrt{8} = \sqrt{4 \cdot 2} = 2\sqrt{2}$

  b. $\sqrt{45} = \sqrt{9 \cdot 5} = 3\sqrt{5}$

  c. $\sqrt{900} = 30$

  d. $\left(\sqrt{3}\right)^2 = 3$

6. a. $\sqrt{28} = \sqrt{4 \cdot 7} = 2\sqrt{7}$

  b. $\sqrt{32} = \sqrt{16 \cdot 2} = 4\sqrt{2}$

  c. $\sqrt{54} = \sqrt{9 \cdot 6} = 3\sqrt{6}$

  d. $\sqrt{200} = \sqrt{100 \cdot 2} = 10\sqrt{2}$

7. a. $\sqrt{\dfrac{9}{16}} = \dfrac{\sqrt{9}}{\sqrt{16}} = \dfrac{3}{4}$

  b. $\sqrt{\dfrac{25}{49}} = \dfrac{\sqrt{25}}{\sqrt{49}} = \dfrac{5}{7}$

  c. $\sqrt{\dfrac{7}{16}} = \dfrac{\sqrt{7}}{\sqrt{16}} = \dfrac{\sqrt{7}}{4}$

  d. $\sqrt{\dfrac{6}{9}} = \dfrac{\sqrt{6}}{\sqrt{9}} = \dfrac{\sqrt{6}}{3}$

8. a. $\sqrt{\dfrac{1}{4}} = \dfrac{\sqrt{1}}{\sqrt{4}} = \dfrac{1}{2}$

  b. $\sqrt{\dfrac{16}{9}} = \dfrac{\sqrt{16}}{\sqrt{9}} = \dfrac{4}{3}$

  c. $\sqrt{\dfrac{5}{36}} = \dfrac{\sqrt{5}}{\sqrt{36}} = \dfrac{\sqrt{5}}{6}$

  d. $\sqrt{\dfrac{3}{16}} = \dfrac{\sqrt{3}}{\sqrt{16}} = \dfrac{\sqrt{3}}{4}$

9. a. $\sqrt{54} \approx 7.35$ and $3\sqrt{6} \approx 7.35$

  b. $\sqrt{\dfrac{5}{16}} \approx 0.56$ and $\dfrac{\sqrt{5}}{4} \approx 0.56$

10. a. $\sqrt{48} \approx 6.93$ and $4\sqrt{3} \approx 6.93$

  b. $\sqrt{\dfrac{7}{9}} \approx 0.88$ and $\dfrac{\sqrt{7}}{3} \approx 0.88$

11. $x^2 - 6x + 8 = 0$
$(x-4)(x-2) = 0$
$x - 4 = 0$ or $x - 2 = 0$
$x = 4$ or $x = 2$

12. $x^2 + 4x = 21$
$x^2 + 4x - 21 = 0$
$(x+7)(x-3) = 0$
$x + 7 = 0$ or $x - 3 = 0$
$x = -7$ or $x = 3$

13. $3x^2 - 51x + 180 = 0$
$3(x^2 - 17x + 60) = 0$
$3(x-12)(x-5) = 0$
$x - 12 = 0$ or $x - 5 = 0$
$x = 12$ or $x = 5$

14. $2x^2 + x - 6 = 0$
$(2x-3)(x+2) = 0$
$2x - 3 = 0$ or $x + 2 = 0$
$2x = 3$ or $x = -2$
$x = \dfrac{3}{2}$ or $x = -2$

15. $3x^2 = 10x + 8$
$3x^2 - 10x - 8 = 0$
$(3x+2)(x-4) = 0$
$3x + 2 = 0$ or $x - 4 = 0$
$3x = -2$ or $x = 4$
$x = -\dfrac{2}{3}$ or $x = 4$

16. $8x^2 + 40x - 112 = 0$
$8(x^2 + 5x - 14) = 0$
$8(x + 7)(x - 2) = 0$
$x + 7 = 0$ or $x - 2 = 0$
$x = -7$ or $x = 2$

17. $6x^2 = 5x - 1$
$6x^2 - 5x + 1 = 0$
$(3x - 1)(2x - 1) = 0$
$3x - 1 = 0$ or $2x - 1 = 0$
$3x = 1$ or $2x = 1$
$x = \dfrac{1}{3}$ or $x = \dfrac{1}{2}$

18. $12x^2 + 10x = 12$
$12x^2 + 10x - 12 = 0$
$2(6x^2 + 5x - 6) = 0$
$2(3x - 2)(2x + 3) = 0$
$3x - 2 = 0$ or $2x + 3 = 0$
$3x = 2$ or $2x = -3$
$x = \dfrac{2}{3}$ or $x = -\dfrac{3}{2}$

19. $x^2 - 7x + 10 = 0$
$a = 1, \ b = -7, \ c = 10$
$x = \dfrac{-b \pm \sqrt{b^2 - 4ac}}{2a}$
$x = \dfrac{7 \pm \sqrt{49 - 4(1)(10)}}{2(1)}$
$x = \dfrac{7 \pm \sqrt{49 - 40}}{2}$
$x = \dfrac{7 \pm \sqrt{9}}{2}$
$x = \dfrac{7 + 3}{2}$ or $x = \dfrac{7 - 3}{2}$
$x = 5$ or 2

20. $x^2 + 7x + 12 = 0$
$a = 1, \ b = 7, \ c = 12$
$x = \dfrac{-b \pm \sqrt{b^2 - 4ac}}{2a}$
$x = \dfrac{-7 \pm \sqrt{49 - 4(1)(12)}}{2(1)}$
$x = \dfrac{-7 \pm \sqrt{49 - 48}}{2}$
$x = \dfrac{-7 \pm \sqrt{1}}{2}$
$x = \dfrac{-7 + 1}{2}$ or $x = \dfrac{-7 - 1}{2}$
$x = -3$ or -4

21. $x^2 + 9 = 7x$
$x^2 - 7x + 9 = 0$
$a = 1, \ b = -7, \ c = 9$
$x = \dfrac{-b \pm \sqrt{b^2 - 4ac}}{2a}$
$x = \dfrac{7 \pm \sqrt{49 - 4(1)(9)}}{2(1)}$
$x = \dfrac{7 \pm \sqrt{49 - 36}}{2}$
$x = \dfrac{7 \pm \sqrt{13}}{2} \approx 5.30$ or 1.70

22. $2x^2 + 3x = 6$
$2x^2 + 3x - 6 = 0$
$a = 2, \ b = 3, \ c = -6$
$x = \dfrac{-b \pm \sqrt{b^2 - 4ac}}{2a}$
$x = \dfrac{-3 \pm \sqrt{9 - 4(2)(-6)}}{2(2)}$
$x = \dfrac{-3 \pm \sqrt{9 + 48}}{4}$
$x = \dfrac{-3 \pm \sqrt{57}}{4} \approx 1.14$ or -2.64

23. $x^2 - 4x - 8 = 0$
$a = 1, \ b = -4, \ c = -8$
$x = \dfrac{-b \pm \sqrt{b^2 - 4ac}}{2a}$
$x = \dfrac{4 \pm \sqrt{16 - 4(1)(-8)}}{2(1)}$
$x = \dfrac{4 \pm \sqrt{16 + 32}}{2}$
$x = \dfrac{4 \pm \sqrt{48}}{2}$
$x = \dfrac{4 \pm 4\sqrt{3}}{2} \approx 5.46$ or -1.46

24. $x^2 - 6x - 2 = 0$
$a = 1, \ b = -6, \ c = -2$
$x = \dfrac{-b \pm \sqrt{b^2 - 4ac}}{2a}$
$x = \dfrac{6 \pm \sqrt{36 - 4(1)(-2)}}{2(1)}$
$x = \dfrac{6 \pm \sqrt{36 + 8}}{2}$
$x = \dfrac{6 \pm \sqrt{44}}{2}$
$x = \dfrac{6 \pm 2\sqrt{11}}{2}$
$x = 3 \pm \sqrt{11} \approx 6.32$ or -0.32

25.
$$5x^2 = 3x + 7$$
$$5x^2 - 3x - 7 = 0$$
$$a = 5, \ b = -3, \ c = -7$$
$$x = \frac{-b \pm \sqrt{b^2 - 4ac}}{2a}$$
$$x = \frac{3 \pm \sqrt{9 - 4(5)(-7)}}{2(5)}$$
$$x = \frac{3 \pm \sqrt{9 + 140}}{10}$$
$$x = \frac{3 \pm \sqrt{149}}{10} \approx 1.52 \text{ or } -0.92$$

26.
$$2x^2 = 8x - 1$$
$$2x^2 - 8x + 1 = 0$$
$$a = 2, \ b = -8, \ c = 1$$
$$x = \frac{-b \pm \sqrt{b^2 - 4ac}}{2a}$$
$$x = \frac{8 \pm \sqrt{64 - 4(2)(1)}}{2(2)}$$
$$x = \frac{8 \pm \sqrt{64 - 8}}{4}$$
$$x = \frac{8 \pm \sqrt{56}}{4}$$
$$x = \frac{8 \pm 2\sqrt{14}}{4}$$
$$x = \frac{4 \pm \sqrt{14}}{2} \approx 3.87 \text{ or } 0.13$$

27.
$$2x^2 = 14$$
$$x^2 = 7$$
$$x = \pm\sqrt{7}$$
$$x \approx \pm 2.65$$

28.
$$2x^2 = 14x$$
$$2x^2 - 14x = 0$$
$$2x(x - 7) = 0$$
$$2x = 0 \quad \text{or} \quad x - 7 = 0$$
$$x = 0 \quad \text{or} \quad x = 7$$

29.
$$4x^2 - 25 = 0$$
$$4x^2 = 25$$
$$x^2 = \frac{25}{4}$$
$$x = \pm\frac{5}{2}$$

30.
$$4x^2 - 25x = 0$$
$$x(4x - 25) = 0$$
$$x = 0 \quad \text{or} \quad 4x - 25 = 0$$
$$x = 0 \quad \text{or} \quad 4x = 25$$
$$x = 0 \quad \text{or} \quad x = \frac{25}{4}$$

31.
$$ax^2 - bx = 0$$
$$x(ax - b) = 0$$
$$x = 0 \quad \text{or} \quad ax - b = 0$$
$$x = 0 \quad \text{or} \quad ax = b$$
$$x = 0 \quad \text{or} \quad x = \frac{b}{a}$$

32.
$$ax^2 - b = 0$$
$$ax^2 = b$$
$$x^2 = \frac{b}{a}$$
$$x = \pm\sqrt{\frac{b}{a}}$$
$$x = \pm\frac{\sqrt{ab}}{a}$$

33. Let the length $= x + 3$ and width $= x$. The area is then:
$$x(x + 3) = 40$$
$$x^2 + 3x = 40$$
$$x^2 + 3x - 40 = 0$$
$$(x + 8)(x - 5) = 0$$
$$x + 8 = 0 \quad \text{or} \quad x - 5 = 0$$
$$x = -8 \quad \text{or} \quad x = 5$$

Reject $x = -8$ because the length cannot be negative. The rectangle is 5 by 8.

34.
$$x \cdot (x + 5) = (x + 1) \cdot 4$$
$$x^2 + 5x = 4x + 4$$
$$x^2 + 1x - 4 = 0$$
$$a = 1, \ b = 1, \ c = -4$$
$$x = \frac{-b \pm \sqrt{b^2 - 4ac}}{2a}$$
$$x = \frac{-1 \pm \sqrt{1 - 4(1)(-4)}}{2(1)}$$
$$x = \frac{-1 \pm \sqrt{1 + 16}}{2}$$
$$x = \frac{-1 \pm \sqrt{17}}{2}$$
$$CP = \frac{-1 + \sqrt{17}}{2} \approx 1.56$$

$\dfrac{-1 - \sqrt{17}}{2}$ is rejected because it is a negative number.

35.
$$D = \frac{n(n - 3)}{2}$$
$$9 = \frac{n(n - 3)}{2}$$
$$18 = n^2 - 3n$$
$$0 = n^2 - 3n - 18$$
$$0 = (n - 6)(n - 3)$$
$$n = 6 \text{ or } n = -3$$
$$n = 6 \text{ reject } n = -3.$$

36. $D = \dfrac{n(n-3)}{2}$

$\quad\;\; n = \dfrac{n(n-3)}{2}$

$\quad\; 2n = n^2 - 3n$

$\quad\;\;\; 0 = n^2 - 5n$

$\quad\;\;\; 0 = n(n-5)$

$\quad n = 0 \;\;$ or $\;\; n - 5 = 0$

$\quad n = 0 \;\;$ or $\quad\;\;\; n = 5$

$\quad n = 5 \;$ reject $n = 0$.

37. $c^2 = a^2 + b^2$

$\quad c^2 = 3^2 + 4^2$

$\quad c^2 = 9 + 16$

$\quad c^2 = 25$

$\quad\;\; c = \pm 5$

$\quad\;\; c = 5;$ reject $c = -5$

38. $\qquad c^2 = a^2 + b^2$

$\qquad 10^2 = 6^2 + b^2$

$\;\; 100 - 36 = b^2$

$\qquad\; 64 = b^2$

$\qquad\;\; b = \pm 8$

$\qquad\;\; b = 8;$ reject $b = -8$